AQA GCSE BIOLOGY

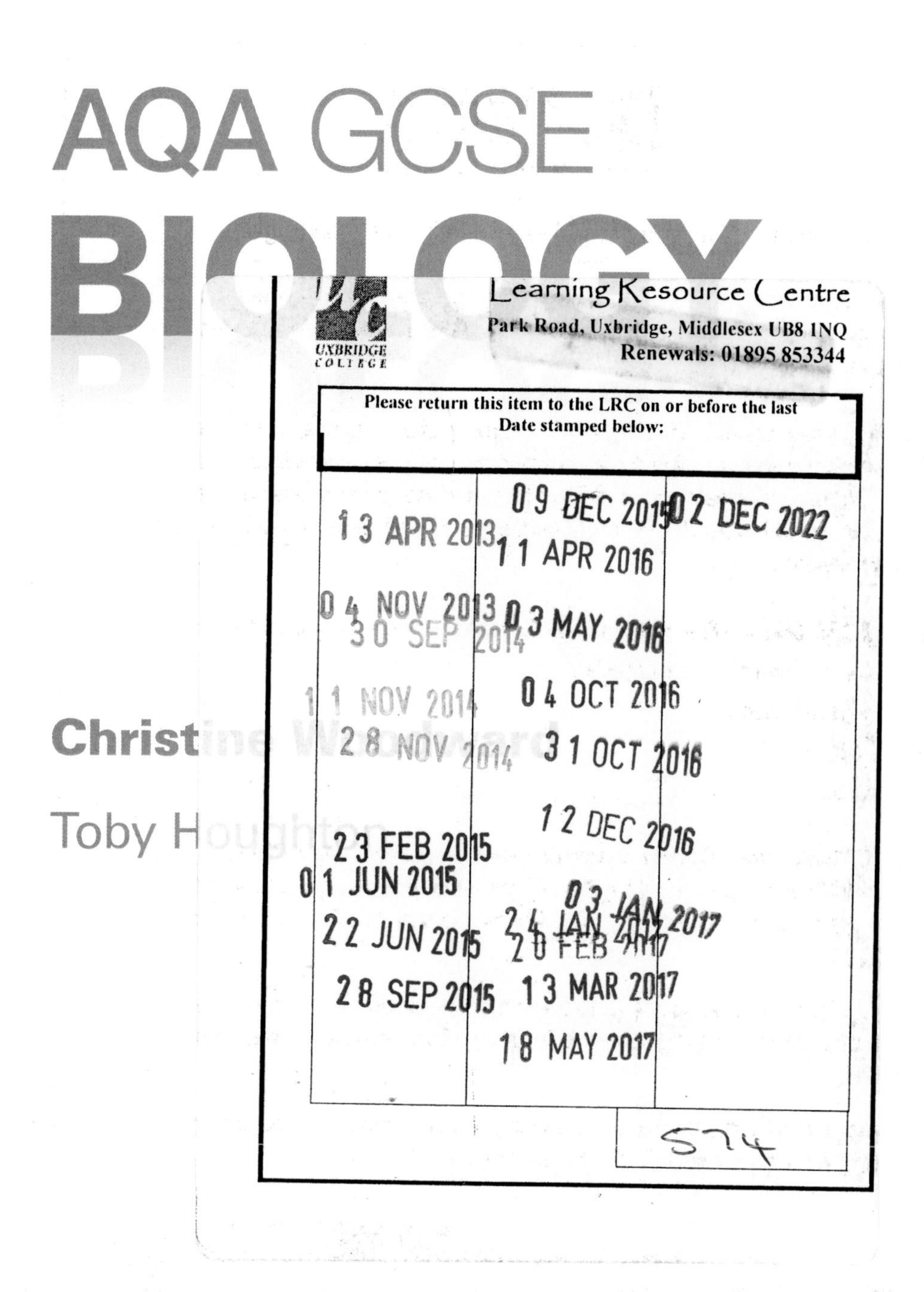

Christine Woodward
Toby Houghton

PHILIP ALLAN UPDATES

Philip Allan Updates, an imprint of Hodder Education, an Hachette UK company, Market Place, Deddington, Oxfordshire OX15 0SE

Orders
Bookpoint Ltd, 130 Milton Park, Abingdon, Oxfordshire OX14 4SB
tel: 01235 827827
fax: 01235 400401
e-mail: education@bookpoint.co.uk

Lines are open 9.00 a.m.–5.00 p.m., Monday to Saturday, with a 24-hour message answering service. You can also order through the Philip Allan Updates website: www.philipallan.co.uk

ISBN 978-1-4441-2080-6

First printed 2010
Impression number 5 4 3
Year 2015 2014 2013 2012

Cover photo: XYZproject/Fotolia

Other photos are reproduced by permission of the following:

p4 Justin Kase z11z/Alamy; **p5** *cl* Tomo Jesenjcnik/Fotolia, *br* Sharpshot/Fotolia; Bob Battersby; **p6** *bl* Maximilian Stock Ltd/Anthony Blake Photo Library, *br* Ingram; **p11** Ingram; **p12** *t* Ronald Hudson/Fotolia, *c* Monkey Business/Fotolia; **p13** TS/Keystone USA/Rex Features; **p14** *c* Scimat/SPL, *b* AMI Images/SPL; **p18** SPL; **p19** Crown copyright (Click-Use PSI licence number C2007001851); **p20** Crown copyright (Click-Use PSI licence number C2007001851); **p.26** Lida Salatian/Fotolia; **p35** Sebastien Maleville/Fotolia; **p36** Martin Shields/SPL; **p37** Martin Shields/SPL; **p38** Nigel Cattlin/SPL; **p39** Colin Cuthbert/SPL; **p43** Tony Marshall/EMPICS–sport; **p45** Martin Shields/SPL; **p46** CuboImages sri/Alamy; **p47** *c* Ilana Vargulich/Fotolia, *b* Oleg Seleznev/Fotolia; **p48** *tc* Christian Musat/Fotolia, *tr* modestlife/Fotolia, *bl* Kim Blackmore/Fotolia; **p49** *tl* Gregory Dimijian/SPL, *tc* Juuliss/Fotolia, *tr* errni/Fotolia, *tc* manfredxy/Fotolia, *b* Frederick R McConnaughey/SPL; **p50** Sylvia Cordaiy Photo Library/Alamy; **p51** *cl* Steve Atkins Photography/Alamy, *cr* Chris Jewiss/Fotolia, *bl* Duncan Noakes/Fotolia; **p52** Christine Woodward; **p54** *cl* Stefan Florea/Fotolia, *bl* Christine Woodward, *br* Jim West/Alamy; **p55** *l* Lea Paterson/SPL, *r* Martin Shields/SPL; **p58** TopFoto; **p59** *tc* Cordelia Molloy/SPL, *tr* Kokhanchikov/Fotolia; **p63** Lucy Cole; **p64** Eric Carr/Alamy; Golden Rice Humanitarian Board (www.goldenrice.org); **p65** *l* Ingram, *c* Ingram, *r* Ingram; **p71** *l* Geoff Kidd/SPL, *bc* Kenneth W Fink/SPL, *br* SCIMAT/SPL; **p74** Shannon Calvert/Connecticut's Beardsley Zoo; **p75** Michael Donne/SPL; **p79** Forgiss/Fotolia; **p81** *tl* carballo/Fotolia, *bl* Gerald & Buff Corsi, Visuals Unlimited/SPL, *tr* Antoine Perroud/Fotolia; **p82** *cr* Marzanna Syncerz/Fotolia, *bl* meryll/Fotolia; **p.83** alessandro merlini/Fotolia; **p84** *tl* Power and Syred/SPL, *tc* Susumu Nishinaga/SPL, *tr* Dr Torsten Wittman/SPL, *bl* Manfred Kage, Peter Arnold Inc. SPL, *br* Steve Gschmeissner/SPL; **p85** Omikron/SPL; **p86** *tl* Dr Gopal Murti/SPL, *cl* Steve Gschmeissner/SPL; **p88** Dr Jeremy Burgess/SPL; **p99** Jeff Carroll/AgstockUSA/SPL; **p100** Biophoto Associates/SPL; **p106** laurent dambies/Fotolia; **p107** *tr* Dennis Cox/Alamy; Krot/Fotolia, *bl* Djordje Korovljevic/Fotolia; Wadsworth Controls; **p112** *cl* Chris Howes/Wild Places Photography/Alamy, *cr* Philippe Psaila/SPL; **p118** Mark Boulton/Alamy; **p119** Chaos/Wikipedia; **p123** Topfoto; **p127** Phototake Inc/Alamy; **p129** Darryl Sleith/Fotolia; **p131** *bl* Ingram, *bc* Ingram, *br* Monkey Business/Fotolia; **p133** Vladimir Wrangel/Fotolia; **p135** *tr* Yuri Arcurs/Fotolia, *cl* John Fryer/Alamy; **p138** Melissa Schakle/Fotolia; **p139** James-King-Holmes/SPL; **p141** Adrian T Sumner/SPL; **p145** *tl* Ingram, *cr* CNRI/SPL; **p150** CNRI/SPL; SPL; **p151** Professor Miodrag Stojkovic/SPL; **p152** Pascal Goetgheluck/SPL; **p155** *cr* Robin Weaver/Alamy, *bl* Lucy Cole; **p156** *tl* Sinclair Stammers/SPL, *cl* Tom McHugh/SPL, *bl* Herve Conge, ISM/SPL, *br* Reuters/Corbis, Reuters/Corbis; **p157** Christian Jegou Publiphoto Diffusion/SPL; **p158** Christine Woodard; **p163** Deborah Benbrook/Fotolia; **p164** Jefery/Fotolia; **p168** TopFoto; **p171** *tl* TopFoto, *tr* Bobjgalindo/Wikimedia; **p175** R. Maisonneuve, Publiphoto Diffusion/SPL; **p179** SPL; **p181** Ralph Hutchings Visuals Unlimited/SPL; **p183** Antonia Reeve/SPL; **p186** Susumu Nishinaga/SPL; **p192** lom123/Fotolia; **p198** picsfive/Fotolia; **p206** SPL; **p207** Everynight Images/Alamy; **p209** *tl* Stephen Finn/Fotolia, *tr* Solar Cookers International (www.solarcookers.org), *b* Walt Parrish/Rotary Club of Fresno; **p210** Justin Kase zsixz/Alamy; **p211** Photoshot Holdings Ltd/Alamy; **p212** Jeff Gynane/Fotolia; **p213** Robert Brook/SPL; **p214** *bl* Picture Contact BV/Alamy, *br* Nancy Sefton/SPL; **p216** BANNER/Fotolia; **p218** Angela Hampton Picture Library/Alamy; **p219** Joe Gough/Fotolia; **p220** Happy Alex/Fotolia; **p222** Professor David Hall/SPL; **p223** Ashden Awards for Sustainable Energy www.ashdenawards.org/Martin Wright; **p224** Ashden Awards for Sustainable Energy www.ashdenawards.org/David Fulford; **p225** *tl* Solar Cookers International (www.solarcookers.org), *br* Martin Bond/SPL; **p226** culture–images GmbH/Alamy; **p228** Bon Appetit/Alamy; **p229** Ashley Whitworth/Fotolia

Printed in Dubai

Hachette UK's policy is to use papers that are natural, renewable and recyclable products and made from wood grown in sustainable forests. The logging and manufacturing processes are expected to conform to the environmental regulations of the country of origin.

P02038

Contents

Getting the most from this book

Welcome to the *AQA GCSE Biology Student's Book*. This book covers all the content of the units Biology 1, Biology 2 and Biology 3 for the new AQA specification, as well as the 'How Science Works' elements. You have to sit three written papers for GCSE Biology plus the Controlled Assessment. The material tested in each written paper matches the content of the three units of this book. The following features are used throughout this book.

Setting the scene

These sections at the start of each chapter introduce the topics that are going to be covered and encourage you to think about what you are going to learn. They also explain how you can develop your **practical** and **ICT** skills through the activities in the chapter.

Learning outcomes

Each section starts with a list of **learning outcomes** that help you to target your thinking. Look back at these outcomes once you have read a section to check that you have understood each of them before moving on.

Boost your grade

Throughout the book there are **Boost your grade** boxes. Each box includes a tip to help you increase your final GCSE grade.

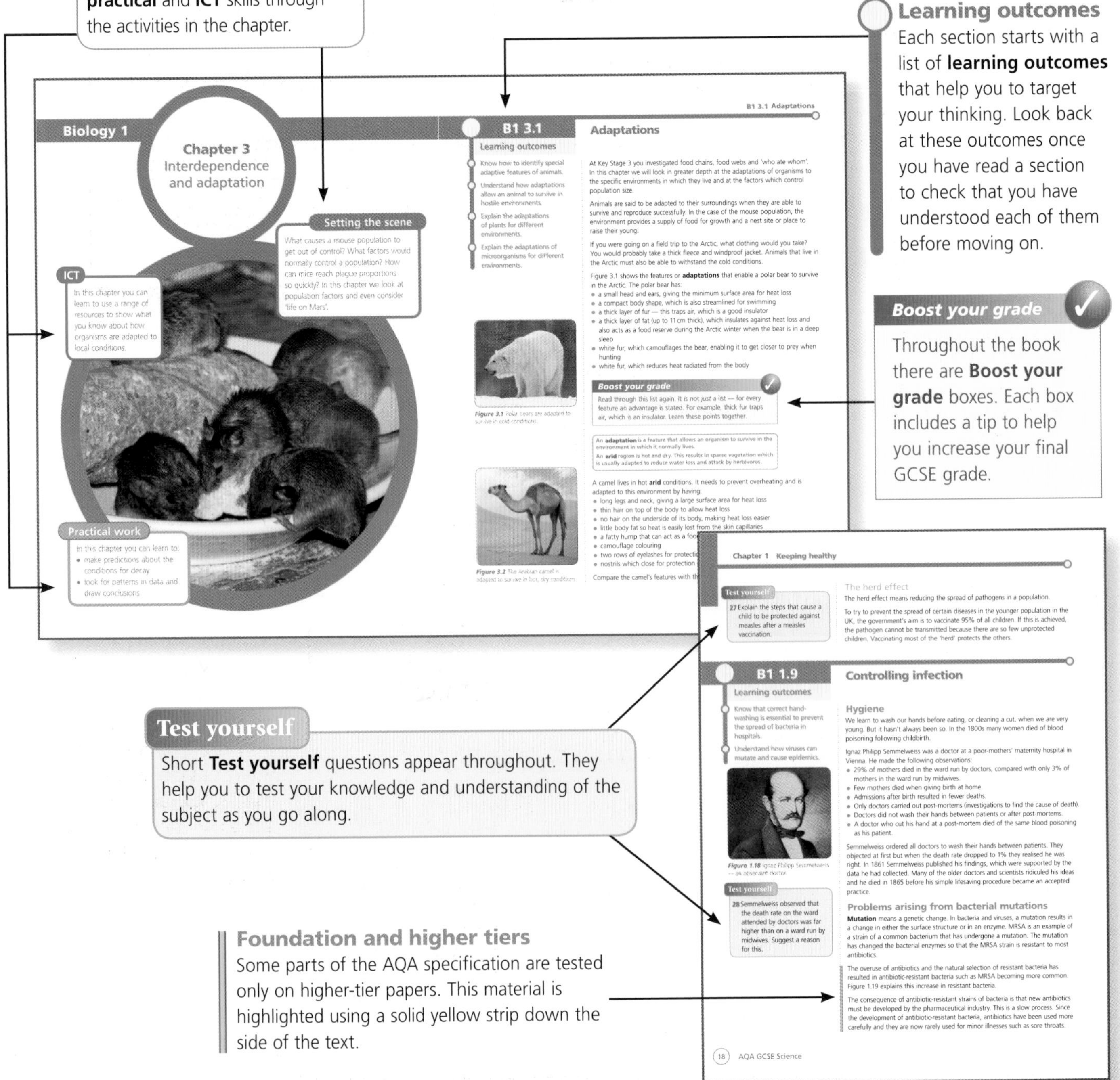

Test yourself

Short **Test yourself** questions appear throughout. They help you to test your knowledge and understanding of the subject as you go along.

Foundation and higher tiers

Some parts of the AQA specification are tested only on higher-tier papers. This material is highlighted using a solid yellow strip down the side of the text.

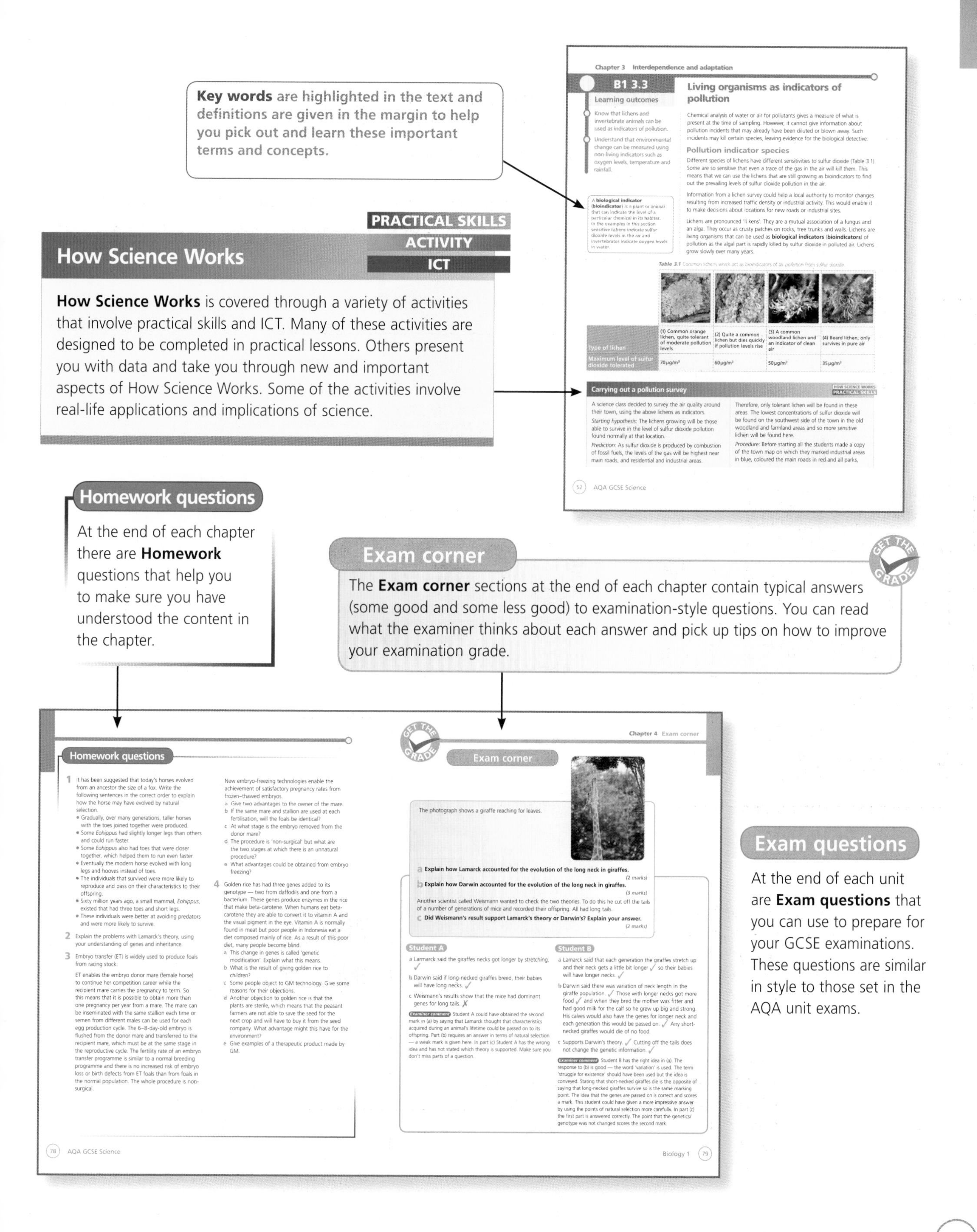
Key words are highlighted in the text and definitions are given in the margin to help you pick out and learn these important terms and concepts.
PRACTICAL SKILLS
ACTIVITY
ICT
How Science Works
How Science Works is covered through a variety of activities that involve practical skills and ICT. Many of these activities are designed to be completed in practical lessons. Others present you with data and take you through new and important aspects of How Science Works. Some of the activities involve real-life applications and implications of science.
Homework questions
At the end of each chapter there are **Homework** questions that help you to make sure you have understood the content in the chapter.
Exam corner
The **Exam corner** sections at the end of each chapter contain typical answers (some good and some less good) to examination-style questions. You can read what the examiner thinks about each answer and pick up tips on how to improve your examination grade.
GET THE GRADE
Exam questions
At the end of each unit are **Exam questions** that you can use to prepare for your GCSE examinations. These questions are similar in style to those set in the AQA unit exams.
Chapter 3 Interdependence and adaptation
B1 3.3
Living organisms as indicators of pollution
Learning outcomes
Pollution indicator species
Carrying out a pollution survey
52 AQA GCSE Science
Homework questions
78 AQA GCSE Science
Chapter 4 Exam corner
Exam corner
Student A
Student B
Biology 1 79

Chapter 1
Keeping healthy

Setting the scene

What do we mean by healthy? Can you tell the healthy people in this photograph by outward appearance only? What types of medical problems can't you see?

ICT

In this chapter you can learn to:
- use a podcast to explain the role of fats in heart attacks
- use multimedia to create an effective information campaign about healthy lifestyles

Practical work

In this chapter you can learn to:
- identify potential errors in an experiment to find energy values of foods
- use aseptic techniques and work safely with bacteria

B1 1.1 The effect of food on health

Learning outcomes

- Know what comprises a balanced healthy diet.
- Explain how being overweight, malnutrition and a deficiency disease can be linked.
- Understand the importance of different food groups for your health.

Do you eat the same foods every day? Do you think about your choice of food? Most people eat a varied diet that includes everything needed to keep their body healthy. The important word in the last sentence is 'varied', because no single food can provide all the essential nutrients your body needs to function efficiently. Functioning efficiently means that the body is able to use the components of food to make vital hormones and enzymes. These hormones and enzymes control the chemical reactions that take place in every cell, to release energy, and for growth and repair. An easy way to be sure that your diet contains all the essential vitamins and minerals is to eat as many different naturally coloured foods as possible. Red tomatoes, green broccoli and purple plums are three examples.

Figure 1.1 Food can be colourful, interesting and provide vital vitamins and mineral ions.

Test yourself

1 a List eight different coloured fruits or vegetables. Try to include all the colours.
  b Try to name all the fruits and vegetables shown in Figure 1.1.

A '**healthy diet**' means eating foods that will supply the body with all the essential nutrients and provide the correct amount of energy for an active lifestyle. The different types of food you need to eat are listed in Table 1.1 on page 6. **Malnutrition** results from not eating a balanced diet. Malnutrition is seen in people who are underweight and people who are overweight. If a person eats a restricted diet for a long time, such as eating only crisps and drinking cola, he/she will be taking in excess energy but will be short of proteins and vitamins. This may show as excess body fat, poor muscle development (protein deficiency), anaemia (iron deficiency) and spotty skin (vitamin C deficiency).

Test yourself

2 Give two health problems a child might start to experience if he or she refuses to eat fruit and vegetables for several weeks.

3 Give three advantages of eating wholemeal bread and brown rice.

4 Eating a chocolate bar and drinking a fizzy drink high in sugar may give you an instant energy fix. List three risks associated with doing this every morning and afternoon break.

Figure 1.2 Wholemeal keeps you feeling fuller for longer. Why?

Figure 1.3 What are the advantages of these foods?

Table 1.1 For a healthy body you should eat food from each food type in column 1 every day — but select carefully.

Food type	Functions	Examples	Advantages/problems
Carbohydrates	Broken down in cells to release chemical energy that fuels the cell's functions Energy for muscle contractions and nerve impulses	Yams, potatoes Whole grains: rice, millet Cereals: wholemeal bread, pasta	Converted slowly to sugars, which maintains a steady blood glucose level Naturally packaged with cellulose fibre, which prevents constipation and keeps the gut healthy, reducing the risk of bowel cancer
		White sugar, high sugar drinks	Quickly converted to fat deposits in the body Causes tooth decay, increases the risk of type 2 diabetes Does not contain any vitamins or mineral ions
Proteins	For making new cells for growth, e.g. building up muscles Repair of damaged cells For making enzymes and hormones that regulate the body	Nuts, tofu, beans, lentils, quorn (mycoprotein)	Vegetable proteins do not contain all the amino acids, so vegetarians must eat a variety of protein foods to ensure they get the essential amino acids
		Meat: red meat, poultry, sausages, bacon, eggs Fish: salmon, sardines, cod	Animal proteins do contain all amino acids Meats contain iron, which is necessary for producing haemoglobin in red blood cells; if this is in short supply, a person will suffer from anaemia
Fats	Protect vital organs from damage Energy reserve Essential fatty acids (EFAs) are used to make hormones, e.g. sex hormones Cholesterol for membranes Fats contain fat-soluble vitamins (A, D, E)	Dairy products and red meats — these foods should be eaten in moderation	Contain saturated fats, which can be used for energy but any excess is stored in the body fat deposits Can raise blood cholesterol level
		Olive oil, peanuts, avocados	Contain monounsaturated fats, which can be used for energy and help to lower blood cholesterol level
		Oily fish, sunflower seeds, vegetable oils	Contain polyunsaturated fats and essential fatty acids, called omega-6 and omega-3, which help to lower the blood cholesterol level
Mineral ions	Calcium ions for strong bones and teeth	Sources of calcium: dairy products — milk, cheese, yoghurt, fromage frais; soya or rice milk with added calcium	Needed in small amounts to keep enzymes functioning
	Iron for making haemoglobin in the blood	Sources of iron: red meat and dark leafy vegetables	
Vitamins	Vitamin C for healthy skin Vitamin A for healthy eyes	Vegetables Salad Fruit Dried fruit	Needed in very small amounts to help cells form enzymes

Figure 1.4 What will these help to reduce?

Figure 1.5 Why should these be eaten in moderation?

Cholesterol

Cholesterol is a word you will hear frequently. It is often made to sound 'bad'. Cholesterol is important in the body for making strong cell membranes and vital hormones. The cholesterol level in the blood depends on:

- the amount produced by the liver (an **inherited factor**)
- the amount eaten in food (an **environmental factor**)

A person eating a balanced diet normally has no problems. Too much fried food, burgers, cheese and pastry can cause a high cholesterol level. This can lead to deposits in the walls of blood vessels, reducing the blood flow and causing high blood pressure.

In some families, an inherited condition means that the liver does not control the cholesterol level and too much cholesterol is produced. This is a medical condition that is treated with cholesterol-reducing medicine.

Test yourself

5 Salmon and mackerel are often referred to as 'brain food' and the advice is to eat these twice a week. In addition to protein, what important nutrients do they contain?

6 A piece of chicken could be grilled or deep-fried. Explain which method would be healthier.

7 Give five reasons why there should be fat in your diet. Name three foods containing fats that are beneficial and three foods that should not be eaten every day.

8 Heart disease is an increasing problem in the UK. What should you control in your diet to reduce your risk of build-up of fatty deposits in artery walls?

9 Haemoglobin in red blood cells transports oxygen. Which mineral ion is needed for production of haemoglobin?

10 Which organ in the body produces cholesterol?

Food energy values

ACTIVITY

Find the food item at home that has the highest value per 100 g for:

- energy
- protein
- carbohydrates
- saturated fats
- unsaturated fats
- salt

Fats and heart attacks

ICT

Create a podcast for the British Heart Foundation explaining the role of fats in heart attacks.

B1 1.2 The energy content of food

Learning outcomes

- Know that different foods release different quantities of energy.
- Evaluate factors that reduce the accuracy of an experiment.

Evaluation of an experiment

HOW SCIENCE WORKS
PRACTICAL SKILLS

Energy values are given on the 'nutritional information' panels on food packets. On some, the old units, calories, are still used. The following experiment will help you to understand how manufacturers obtain the energy value for a food product.

An experiment to compare the energy values of foods might be to take pieces of food, set fire to each in turn, hold a boiling tube containing water above the flame, and see which food heats the water the most.

How many factors have not been controlled in this suggestion and the apparatus shown in Figure 1.6? Think about this before you look at the hints below.

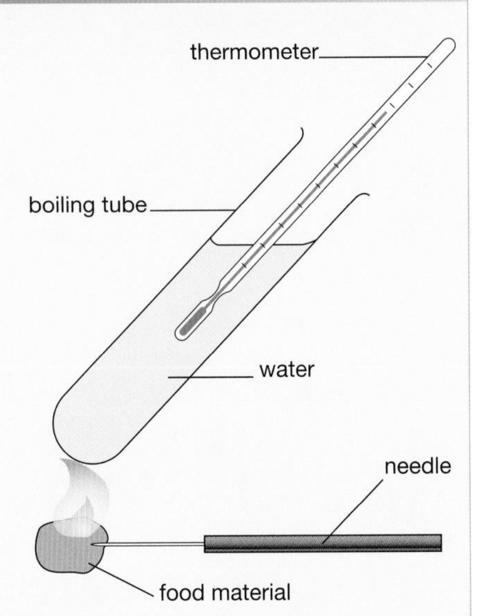

Figure 1.6 Apparatus used to measure the heat energy of various foods in a school laboratory.

Did you consider the following points?

1. There should be only one independent variable and for this experiment we are trying to compare the energy values of different foods. All other possible variables must stay the same (control variables) for each different food. How could you make the mass of food a **fair test**?
2. Have you tried to burn foods? It can take quite a lot of heat to start the reaction in air. If you heated the foods over a Bunsen burner for different times, this would be an additional variable. Can you think of another method of getting the food to burn?
3. Do substances burn better in oxygen than in air?
4. What about the volume of water? Should the water be stirred to distribute the heat?
5. Will all the heat go into the water in a boiling tube? If not, where will it go? Could you reduce heat losses? Why could this be described as a **systematic error**?
6. The boiling tube is made of glass. Is glass the best material to use or could you suggest a better conductor of heat?
7. How would you measure the temperature rise of the water? How would you make sure your results are **accurate**? How could you make them **precise**?

After you have evaluated this experiment, look at the diagram of the food calorimeter in Figure 1.7 and see how many of your criticisms have been overcome.

Systematic errors affect all results in an experiment. The results are shifted away from the true value because of an error or fault in the apparatus.

Zero errors are a type of systematic error, caused by using a measuring instrument that does not read zero before taking the measurement.

Random errors cause readings to vary from the true value. They can be compensated for by taking a large number of readings. The **true value** is the value you would get if you could measure the quantity without any errors at all.

A **fair test** is one in which only the independent variable is allowed to affect the dependent variable. All other factors that could affect the outcome are kept constant.

Accuracy is how close results are to the true value.

Precision is related to how closely values obtained by repeated measurements agree.

Reliability is the ability to repeat the experiment and get similar results.

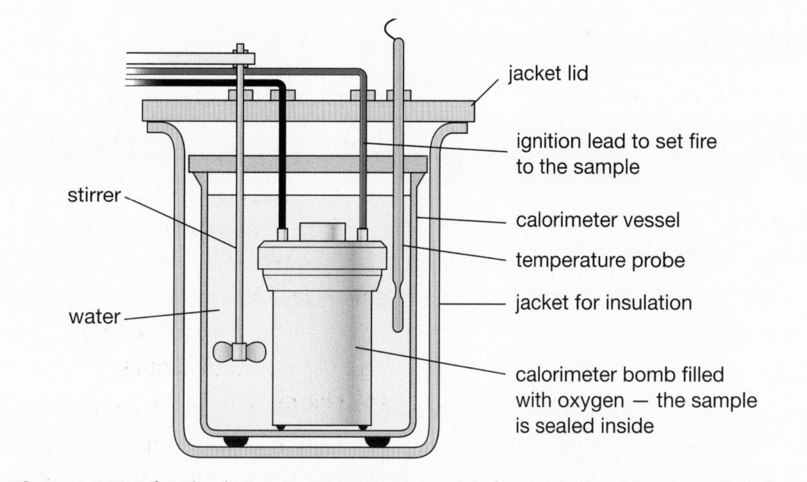

Figure 1.7 A food calorimeter. Experiments with food calorimeters show that 1 gram of fat contains more than twice as much energy as 1 gram of carbohydrate.

Energy in food

HOW SCIENCE WORKS ACTIVITY

Watch the video *Energy contained in foods* and summarise the main errors in the experimental procedure.

B1 1.3 Exercise for health

Learning outcomes

- Know how to calculate the amount of energy used with exercise.
- Understand the link between exercise, food energy and health.

How does the energy content of food relate to everyday activity? Look at Table 1.2. How does walking faster change the energy used?

Table 1.2 The energy requirements for different activities.
The energy a person weighing 60 kg (approx. 9.5 stone) would use in 30 minutes of different activities.

Activity	Energy used in joules (J)
Walking slowly	314
Walking quickly	628
Pushing an electric/petrol lawnmower	700
Cycling at normal speed	750
Aerobics	810
Swimming (steady crawl)	820
Tennis doubles	630
Tennis singles	1000
Running 1 km in 6.2 minutes	1250
Running 1 km in 4.7 minutes	1700

More energy is used if:
- your weight is greater
- you are fit with well-developed muscles

Calculating energy used

HOW SCIENCE WORKS ACTIVITY

The energy used for an activity depends on body mass, rate of working and the duration of exercise. The following example demonstrates this.

1. Copy Table 1.3 and calculate the total energy used by Helen in a typical day. Assume her body mass is 60 kg. Use the data given in Table 1.2

Table 1.3 Helen's activity record

Activity	Time in minutes (min)	(Show your working here)	Energy used (kJ)
Helen cycles to and from school, taking 7.5 minutes each way	Total time 15	Energy used for 30 minutes cycling is 750 kJ Energy used in cycling for 15 minutes = 750 ÷ 2	375
On arrival at school she does 1 hour of swimming training.			
At lunchtime she walks around slowly while chatting	15		
In the evening she plays a game of tennis (doubles) for 1 hour			
		Total	

2. Copy Table 1.4 and calculate the total energy used by Saroj in a typical day. Assume that his body mass is 60 kg.

Table 1.4 Saroj's activity record

Activity	Time (min)	(Show your working here)	Energy used (kJ)
Saroj starts the day by doing a paper round. This takes him 30 minutes walking quickly			
He is late so he runs the 2.0 km to school, getting there in 12 minutes			
He rests at lunchtime		0	0
He walks home slowly, taking 1 hour to get there			
In the evening he earns more pocket money by mowing an elderly person's lawn. He takes 45 minutes using an electric push mower			
		Total	

3. How would you use these calculations when giving advice to Helen and Saroj on what they should eat?

Test yourself

11 What improvements to the heart, blood vessels and blood composition would you expect after starting (and keeping up) regular fitness training?

12 'Take the stairs not the lift'. Give three advantages this would have for a female office worker's muscles, tendons and bones.

13 After 6 weeks of training you are told that your metabolic rate will have increased. Why might this mean that you are both gaining muscle and losing weight?

The benefits of exercise

Exercise is good for you because it:

- increases the strength of your heart muscles and reduces blood pressure
- improves lung ventilation (more air is moved in and out of the lungs)
- improves oxygen uptake into the blood, which benefits your brain and body
- increases muscle size and efficiency
- strengthens ligaments and tendons and reduces the risk of injury
- increases bone density and reduces the risk of osteoporosis (bone wastage)
- decreases cholesterol levels in the blood and reduces the risk of blocked blood vessels
- improves resistance to infections
- raises metabolic rate and continues 'to burn up food' after you have finished exercising
- leaves you with a 'feel good factor'

If you check back, you can see that exercise and diet work together on improving the health of the whole body.

ACTIVITY

You are a nutritional advisor to your country's Olympic team. Create a suitable weekly diet for:

1. a shot putter
2. a marathon runner

B1 1.4

Learning outcomes

- Know why people who exercise regularly are usually healthier than people who do not.
- Understand the term metabolic rate and how this is increased by regular exercise.

Metabolic rate is the rate at which food is broken down chemically in our bodies. It can be measured in joules per minute.

Environment or genetics?

A body's **metabolic rate** is affected by genetic (inherited) and environmental (lifestyle) factors. Lifestyle factors include what we eat and the type and duration of exercise we take. Humans evolved as active beings who were able to survive in an environment where food was sometimes in short supply. Our ancestors were hunter-gatherers and farmers. They had a good **energy in/energy used balance**. There are rarely any food shortages in our supermarkets today, and our journey there by car does not use up much of our energy intake. So compared with our ancestors we have a double

Figure 1.8 How many of these modern weight traps do you fall for? Suggest healthy alternatives.

Figure 1.9 Regular exercise helps to balance the energy in and energy used.

disadvantage — more energy (food) in and less energy used (exercise) in our daily lives. The consequence of this is gain in weight.

To balance the amount of energy in with the amount of energy used and prevent weight gain we should:
- take regular exercise
- be aware that fried and fatty foods contain more energy
- know which foods contain hidden fats and sugars that provide surplus energy

Muscle tissue has a higher metabolic rate than fatty tissue, so a muscular person uses up more energy than a skinny or fat person. Spending leisure hours playing computer games or watching television and DVDs does not use energy or build up muscles. Metabolic rate may also be affected by inherited factors.

Test yourself

14 It is estimated that children today consume the same food energy as children did 20 years ago. Suggest three reasons why many children are fatter today than in the past.

15 The number of overweight children suffering from type 2 diabetes is increasing. Refer to Table 1.2 to explain why exercise in addition to a balanced diet will help them.

Food, health and exercise

HOW SCIENCE WORKS ACTIVITY

1. Working as a group, think through a typical day and list:
   a the high-energy foods you ate or avoided
   b the five portions of fruit and vegetables you ate
   c the activity opportunities you took or avoided
2. Is your lifestyle healthy or could you still make some positive changes for improved health? Remember, any changes must be fun if you are going to keep them up.
3. Doctors say that small changes, such as walking up stairs, taking a slightly longer route to walk home and using a bike instead of getting a lift, are more beneficial than big ideas such as 'I'm going to jog 5 miles three times a week'. Suggest some reasons why.
4. Produce a brief set of recommendations from the group for a healthy lifestyle for teenagers.

Figure 1.10 Be aware of what you are eating. What are the risks here?

B1 1.5

Learning outcomes

- Know the effect of food on health.
- Be able to combine all you have studied about food and exercise to work out a healthy lifestyle.
- Evaluate claims made by slimming programmes and products.

What happens when the diet is not balanced?

The fact is that if energy in the food you eat is greater than the energy you use, you will put on weight. The calculations in Tables 1.3 and 1.4 show a direct link between exercise and energy used. The key facts are:
- The less exercise you take, the less food you need.
- If you take in more energy than you use, the excess is stored as fat.

Obesity is a global health issue. The World Health Organization (WHO) estimates that there are 1 billion overweight adults, of whom 300 million are **obese**. Obesity and the health problems it causes used to be problems of the 'over 50s'. As well as being subject to unkind name-calling, overweight people are more likely to suffer from arthritis, type 2 diabetes, high blood pressure and heart disease than thinner people. Some diseases such as diabetes and rising blood pressure are even seen in young teenagers who are overweight.

If the energy taken in is less than the energy needed for daily activity, this can also cause problems. You have probably heard of anorexia — a condition in which someone controls their food intake well below that needed for health. Anorexia affects about 1% of females. It is the third most common long-term illness in teenagers, with a high mortality rate. Sufferers are tired, short of energy, their skin is damaged easily, they feel cold and have no menstrual cycle. If the problem continues, they may develop an irregular heartbeat, which can lead to heart failure, and loss of bone mass can result in osteoporosis at a young age.

Malnutrition occurs when there are food shortages caused by drought, famine or flooding in developing countries. The people suffer deficiency diseases as a result of shortages of protein, vitamins, minerals and essential amino and fatty acids (see Section 1.1). They have low resistance to infection and suffer muscle wastage, resulting in low weight and poor development.

Test yourself

16 What diseases previously associated with pensioners are now being seen in teenagers? Why are teenagers suffering from these diseases?

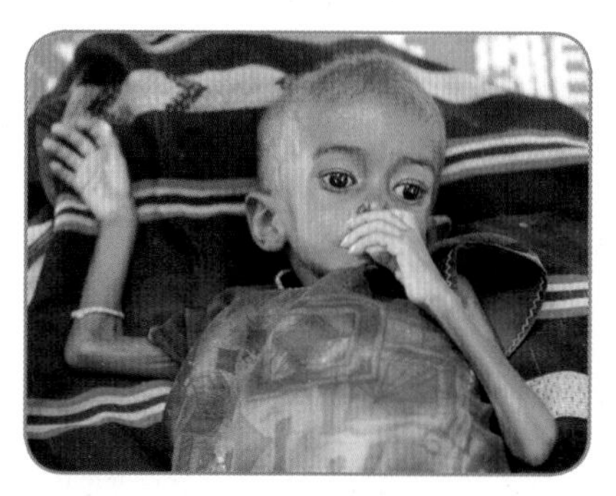

Figure 1.11 This toddler is suffering from protein energy malnutrition. The child has muscle wastage, no fat beneath the skin and is weak and lethargic.

Slimming

Slimmers try to reduce their energy intake so that it is less than the energy used. Their aim is to lose weight and have an attractive well-toned figure. The best way to achieve this is to eat wisely and to exercise more in order to make firm muscles but lose fat.

Evaluating slimming programmes

HOW SCIENCE WORKS ACTIVITY

In groups:

1. Collect three different slimming guides from magazines or books.
2. Evaluate these diets carefully. Use the following questions to produce a tick chart:
   - Will all the foods groups listed in Table 1.1 be eaten each day?
   - Will the essential fatty acids be eaten?
   - Will the diet provide five servings of fruit and vegetables daily?
   - Are the foods in the diet colourful (providing vitamins and minerals)?
   - Are there enough high-fibre foods or will you always feel hungry?
   - Is an exercise routine included?
   - Is the exercise routine feasible and one you will stick with?
   - Are there any comments you would like to add?
3. Present your findings to the rest of the class in an interesting way.

Check your lifestyle

HOW SCIENCE WORKS ICT

Create a resource using ICT that will allow people to check their lifestyle in terms of diet and activity level to see how healthy they are and what they need to do to become healthier. You could create a website, presentation, leaflet, podcast or video.

Test yourself

17 You missed breakfast. You had a mid-morning chocolate bar and a bottle of high-sugar drink. You felt hungry again an hour later. Why would a wholemeal sandwich have been better (a) in the short term and (b) in the longer term if this was a regular eating pattern?

18 A diet consists of eating only eggs. Suggest some problems a dieter might experience if this diet were followed for several weeks.

B1 1.6 What causes infectious disease?

Learning outcomes

- Know that many diseases are caused by pathogens.
- Understand that antibiotics kill bacteria.

Pathogens are microorganisms (bacteria or viruses) that cause disease.

We have seen that what we eat and how active we are can affect our health. Now let's look at the topic of disease.

Infectious diseases are caused by **pathogens**. Pathogens are microorganisms, which include bacteria, viruses and fungi. They are called microorganisms because they are very small and can only be seen with a microscope. Different pathogens invade different parts of the body. This helps doctors to identify which disease we have caught.

Bacteria

Bacteria are much larger than viruses but can only be seen using a microscope. Pathogenic bacteria:

- live in parts of the body such as the nose and throat but not inside cells
- reproduce very quickly
- may produce waste products that act as toxins and irritate cell membranes. These can cause sore throats and runny noses

Test yourself

19 How can bacteria cause sore throats?

Antibiotics

Antibiotics are medicines that kill bacteria inside the body that would otherwise cause further illness.

Antibiotics help to cure bacterial infections by killing bacteria inside the body. The discovery of antibiotics is one of the most important advances in modern medicine. Development occurred during the 1950s. The first antibiotics were used to prevent serious infections from injuries that left open wounds. Antibiotics saved the lives of many children who suffered from bacterial pneumonia, diphtheria and tuberculosis, which were common in overcrowded working and housing conditions.

Different antibiotics attack bacteria in different ways. The most common antibiotic is penicillin. This makes the bacterial cell wall weaker and porous so that the bacteria burst. Other antibiotics prevent the bacteria from reproducing and some interfere with bacterial enzymes.

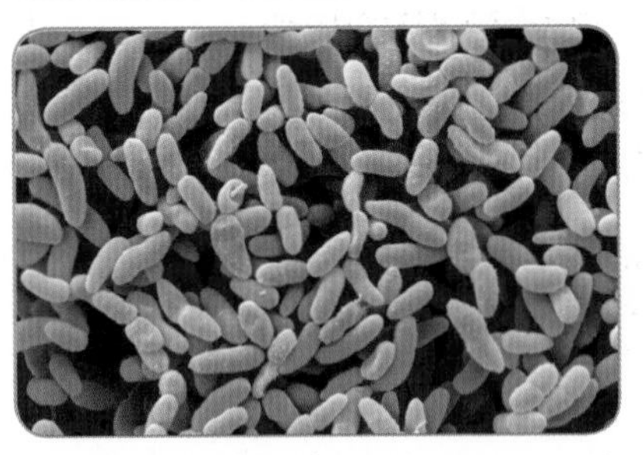

Figure 1.12 These bacteria are seen using a high magnification scanning electron microscope.

Viruses

Viruses are the smallest microorganisms. They:

- are taken into cells of the body
- use the cell contents to replicate and rapidly form thousands of identical copies
- damage the cell as they burst out
- infect nearby cells and repeat the process

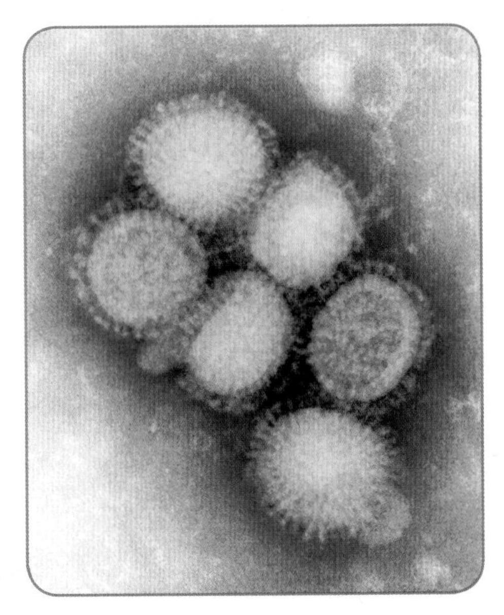

Figure 1.13 These are H1N1 swine flu virus particles. This type can affect humans as well as pigs. It could cause a worldwide outbreak of flu.

Test yourself

20 Which are larger, bacteria or viruses?

21 Why is it difficult to produce a medicine to cure viral infections?

22 Bacteria live on the membranes inside the body. Where in the body do viruses attack?

Replicate means 'form an exact copy'. As viruses are not able to live independently, they are not classed as 'living' so we do not use the word reproduce. They replicate using the DNA of the host cell. The damage to the host cell causes it to die.

Viruses cannot be destroyed using antibiotics. Viruses live and **replicate** inside cells. It is therefore difficult to develop drugs that can enter cells and stop the virus replicating without damaging our body cells. We have to rely on our immune system to fight the virus (see Section 1.7).

B1 1.7

Learning outcomes

- Explain the body's own defences against pathogens.
- Explain how a person becomes immune to a virus infection.

Antibodies are special proteins produced by white blood cells. Antibodies destroy one type of pathogen.

Antigens are protein molecules on the surface of pathogens that trigger an immune response.

The body's natural defence against pathogens

The body has various ways of protecting itself against pathogens — from the effective barrier of the skin to the work of specialised white blood cells.

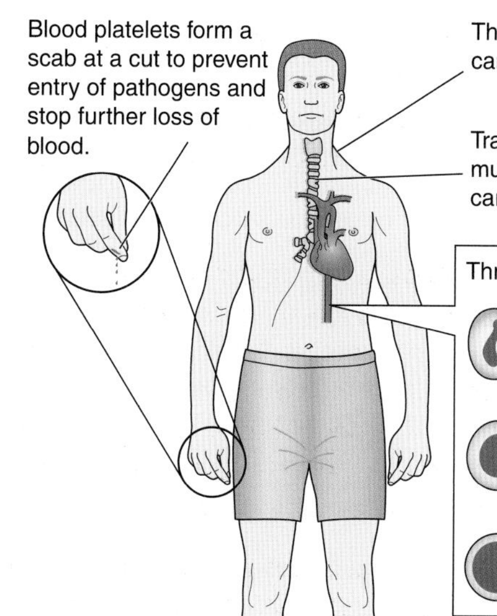

Figure 1.14 The body's natural defences against disease.

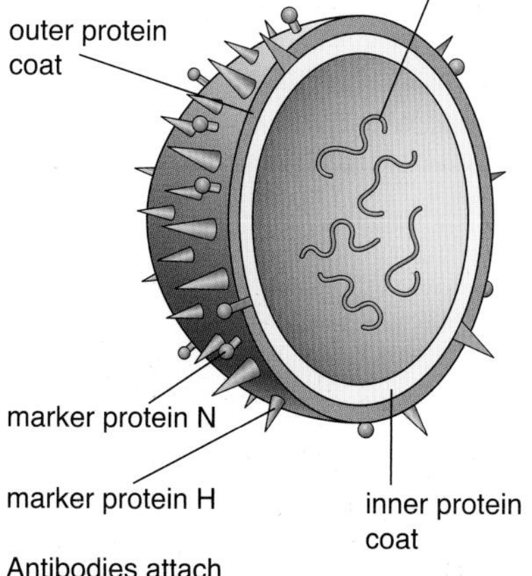

Figure 1.15 This diagram of a virus shows the surface proteins to which antibodies attach.

The role of white blood cells

All pathogens have unique marker proteins called **antigens** on their surface.

When a person catches an infection, their white blood cells respond in one of three ways:

- antibody production
- ingestion of the pathogen
- antitoxin production

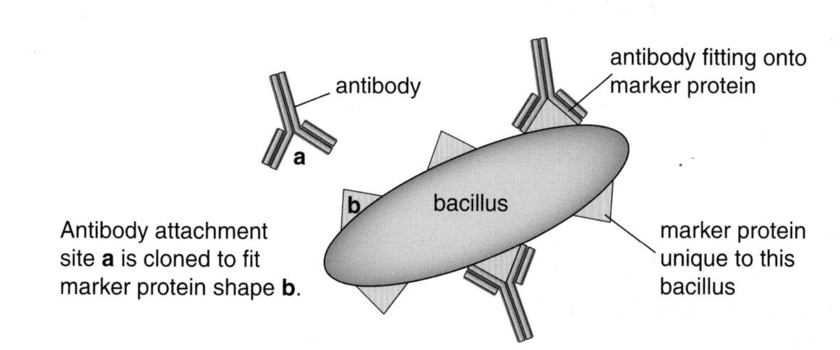

Figure 1.16 How an antibody fits onto the protein marker of the pathogen.

Antibody production

White cells react to pathogens by producing protein antibodies specific to the markers (antigens) on the pathogen surface. (Specific means that the antibody has a structure that fits only one antigen, so there are as many different-shaped antibodies as there are different antigens.)

Antibody production takes a few days. The antibodies fit onto the surface of the pathogen and cause several pathogens to stick together. These groups of pathogens are then ingested by large white cells with a lobed nucleus — see below. Once the antibodies have been produced in sufficient numbers to destroy the pathogens, the patient begins to get better. 'Memory cells' are also produced. These last a long time so if the same pathogen invades the body again years later, memory cells rapidly activate antibody production. The person is said to be immune to that pathogen. Two important points about immunity are (1) that the memory cells enable large numbers of antibodies to be produced and (2) the antibodies are produced at a rate faster than the pathogen replicates.

Boost your grade

Words that sound alike are often muddled and this loses many marks in exams. Remember: anti**bodies** are produced inside the **body** and function inside the **body**.

Do not confuse antibiotics with antibodies. Antibiotics are medicines.

Worked example

How immunity works

- Jane catches chickenpox. Her white cells are activated to produce chickenpox antibodies.
- Chickenpox antibodies attack and destroy the chickenpox virus in her body.
- Memory cells are also produced, which are specific to the chickenpox virus.
- A year later, Jane's friend Sally catches chickenpox and the two girls spend time together watching videos. So Jane comes into contact with the chickenpox virus again.
- Jane's memory cells are activated and her white cells **quickly** produce **large numbers** of chickenpox antibodies, which destroy the virus before it can make her ill. This means she will not suffer from chickenpox twice. We say that Jane is immune to chickenpox.

Test yourself

23 How does an antibody 'fit' onto a pathogen?

24 After the antibodies have caused the pathogens to stick together in a cluster, how are they destroyed?

25 When a person develops an infection, why does it take a few days before he or she starts to get better?

26 Why is a small child unlikely to catch chickenpox a second time?

Ingesting the pathogen

Once the pathogens have been stuck together by antibodies, large white cells with a lobed nucleus ingest the pathogens, taking them into a vacuole. The pathogens are broken down chemically by enzymes, which are added into the vacuole from the white cell cytoplasm.

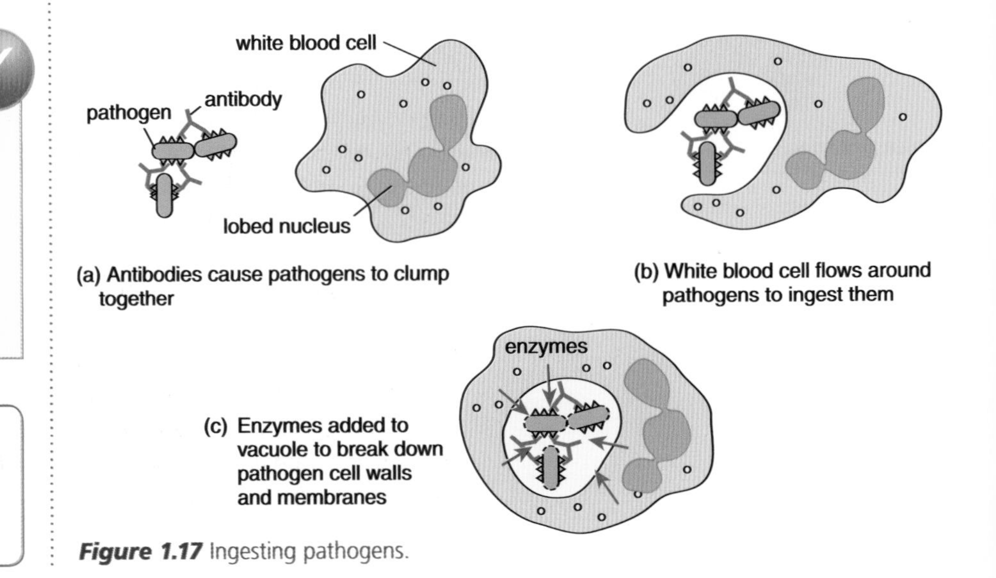

Figure 1.17 Ingesting pathogens.

Boost your grade

This topic is tested frequently and answers are often too vague. Go through the example that explains immunity. Make sure you understand the words 'specific', 'pathogen' and 'antibody'.

Ingest means that the white cell moves around the pathogen and takes it into the cell cytoplasm surrounded by a bit of the cell membrane.

Antitoxin production

Some pathogens produce waste products that act as toxins and irritate membranes (for example, cause sore throats). Specialised white cells produce antitoxins that bind to the toxins to neutralise them. The antitoxins also activate the immune system to destroy the pathogen.

The immune system is fascinating and complex. It has too many cells and chemicals to mention here, all of which work together very effectively to prevent pathogens causing us too much harm. As medical science has progressed, doctors have found ways of artificially helping the body protect against some infectious diseases.

B1 1.8

Learning outcomes

- Understand the importance of immunisation programmes.
- Be able to assess the value of, and problems associated with, antibiotics.

Protecting our bodies against pathogens by vaccination

Our bodies can be protected against some pathogens by **immunisation**, which results in artificial immunity. Immunisation involves injecting a **vaccine**, which contains weak, inactive or dead forms of a pathogen. The vaccine stimulates the white blood cells to produce antibodies specific to the pathogen. Memory cells are also made, which make the person immune to further infection by the same pathogen. In other words, the body is made to respond to the pathogen in the vaccine in the same way as it would to catching an infection. In this way we are prevented from catching the illness.

A **vaccine** is a solution that contains weak, inactive or dead pathogens.

Vaccination can be used to protect children against viruses such as measles, mumps and rubella (MMR). When viral vaccines are prepared, the viral genetic material is removed and only the viral coat is used in the vaccine (see Figure 1.16) but this is sufficient for specific antibodies to be made. A vaccination gives the child lasting immunity because the memory cells activate antibody production on contact with the pathogen. The rate of antibody production means that the pathogen is destroyed before its numbers are sufficient to cause illness.

Vaccination programmes

In developed countries, deaths from infectious diseases are rare because of **vaccination programmes**. Vaccination has been very successful in the UK for the following reasons:

- There is a schedule for vaccination of babies and children up to 15 years of age.
- The majority of children are vaccinated.
- Parents receive computer-printed reminders for booster injections.
- Good supplies of safe vaccine are paid for by the government.
- Refrigerated lorries transport the vaccine.
- Local community nurses carry out the programme.
- Most children are healthy and well fed so their immune systems respond by producing antibodies and memory cells.

A **vaccination programme** is a schedule of injections given to children to prevent common childhood illnesses.

Test yourself

27 Explain the steps that cause a child to be protected against measles after a measles vaccination.

The herd effect

The herd effect reduces the spread of pathogens in a population.

To try to prevent the spread of certain diseases in the younger population in the UK, the government's aim is to vaccinate 95% of all children. If this is achieved, the pathogen cannot be transmitted because there are so few unprotected children. Vaccinating most of the 'herd' protects the others.

B1 1.9 Controlling infection

Learning outcomes

- Know that correct hand-washing is essential to prevent the spread of bacteria in hospitals.
- Understand how viruses can mutate and cause epidemics.

Figure 1.18 Ignaz Philipp Semmelweiss — an observant doctor.

Test yourself

28 Semmelweiss observed that the death rate on the ward attended by doctors was far higher than on a ward run by midwives. Suggest a reason for this.

Hygiene

We learn to wash our hands before eating, or cleaning a cut, when we are very young. But it hasn't always been so. In the 1800s many women died of blood poisoning following childbirth.

Ignaz Philipp Semmelweiss was a doctor at a poor-mothers' maternity hospital in Vienna. He made the following observations:

- 29% of mothers died in the ward run by doctors, compared with only 3% of mothers in the ward run by midwives.
- Few mothers died when giving birth at home.
- Admissions after birth resulted in fewer deaths.
- Only doctors carried out post-mortems (investigations to find the cause of death).
- Doctors did not wash their hands between patients or after post-mortems.
- A doctor who cut his hand at a post-mortem died of the same blood poisoning as his patient.

Semmelweiss ordered all doctors to wash their hands between patients. They objected at first but when the death rate dropped to 1% they realised he was right. In 1861 Semmelweiss published his findings, which were supported by the data he had collected. Many of the older doctors and scientists ridiculed his ideas and he died in 1865 before his simple lifesaving procedure became an accepted practice.

Problems arising from bacterial mutations

Mutation means a genetic change. In bacteria and viruses, a mutation results in a change in either the surface structure or in an enzyme. MRSA is an example of a strain of a common bacterium that has undergone a mutation. The mutation has changed the bacterial enzymes so that the MRSA strain is resistant to most antibiotics.

The overuse of antibiotics and the natural selection of resistant bacteria has resulted in antibiotic-resistant bacteria such as MRSA becoming more common. Figure 1.19 explains this increase in resistant bacteria.

The consequence of antibiotic-resistant strains of bacteria is that new antibiotics must be developed by the pharmaceutical industry. This is a slow process. Since the development of antibiotic-resistant bacteria, antibiotics have been used more carefully and they are now rarely used for minor illnesses such as sore throats.

The less often antibiotics are used, the more likely they are to be effective because the bacterium has had less chance to develop resistance.

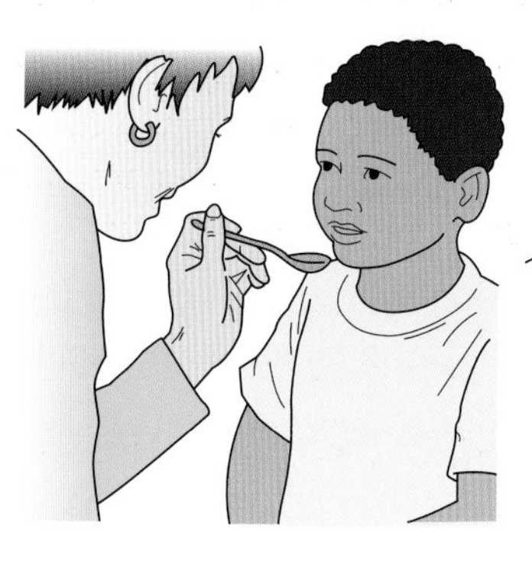
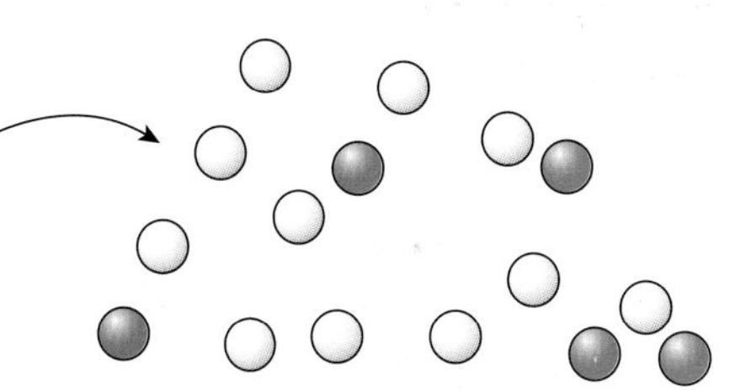

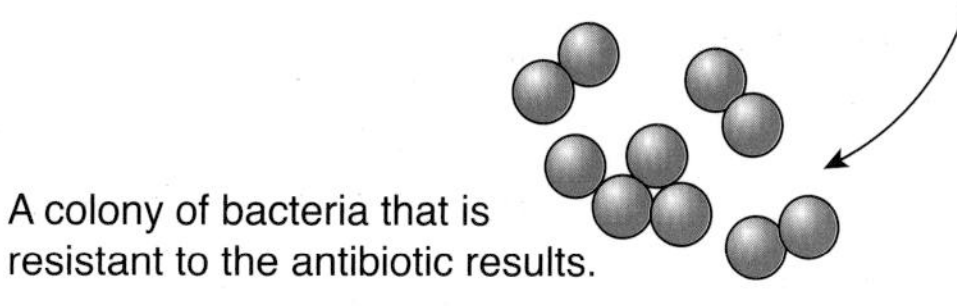

Figure 1.19 Development of a strain of bacteria resistant to antibiotics.

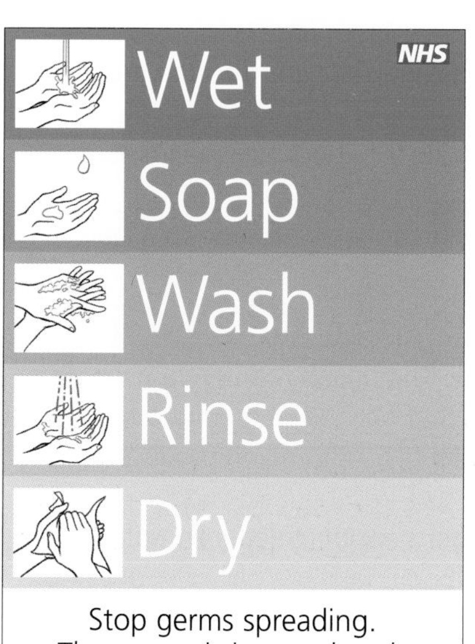

Figure 1.20 A poster showing correct hand-washing procedure.

Epidemic: an infectious disease that spreads to many people in one location, such as a school or a town.

Pandemic: an infectious disease that spreads quickly to many people in many countries, or spreads worldwide. For example, swine flu was identified in Mexico in April 2009 and by June 2009 had spread to 74 countries with at least 30 000 cases.

Still a twenty-first century problem

Just as Semmelweiss had shown for controlling infection in 1861, the best defence against antibiotic-resistant bacteria such as MRSA is careful hand-washing. This has led to a major campaign in all hospitals, with posters reminding staff and visitors to wash their hands or use alcohol gel before entering wards or touching patients. Hospital hygiene has improved but needs constant attention to detail as bacteria are easily spread. Patients who become infected with MRSA are either isolated or barrier nursed. Barrier nursing means that the doctor or nurse wears a disposable apron and gloves whenever they touch the patient which are thrown away afterwards. They must also wash their hands and use alcohol gel before visiting the next patient. This procedure has reduced hospital MRSA infections considerably but unfortunately this is not the only antibiotic-resistant bacterium. Antibiotic-resistant tuberculosis is spreading through the world.

Viruses, rapid mutations and pandemics

Let us start with a quick recap of important points before we see what has happened in some pandemics.

Immunity results when memory cells 'recognise' the shapes of the surface proteins of pathogens. The memory cells rapidly form antibodies to fit onto and destroy the pathogen.

As you can see in Figure 1.15, viruses have a simple structure. The letters H and N refer to the surface proteins. Flu viruses mutate rapidly. This means that spontaneous gene changes result in shape changes in the H and N surface proteins and so new strains are formed. There are many different shapes and combinations and a different antibody has to be made for each. Flu viruses can swap genes between the different strains that infect birds, pigs and humans. This enables the virus to cross species.

Figure 1.21 Why was there such widespread concern about a possible swine flu pandemic in 2009?

In 2005 many people were worried about 'bird flu'. Bird flu is not a new disease. It is **endemic** (commonly found) in waterfowl populations. The new virus strain is mostly in Asia. Bird flu is spread around the world when thousands of ducks and geese migrate to wetland winter-feeding grounds in Britain and other parts of Europe. The virus is excreted in the droppings of infected birds, so it could pass to domestic or farmed geese, ducks, turkeys and chickens. The virus does not spread easily from birds to humans — where humans have caught bird flu it has been by handling infected poultry. All human cases are monitored carefully and so far no human-to-human transmission has been confirmed.

The anxiety about bird flu resulted from:

- the economic consequences if poultry in intensive rearing units were infected and had to be destroyed
- the rapid spread of the virus among the birds
- memories of the serious flu pandemic in 1918 that killed 50 million people worldwide. Recent studies of the genetics show that it was a bird flu virus that mutated to form a completely new strain, which infected both humans and pigs: neither had immunity to this new strain.

Where does the 2009 'swine flu' pandemic fit in?

- The new virus strain has formed from two swine flu viruses that also carry genes from bird and human flu viruses.
- It passes from human to human by the inhalation of virus-infected droplets produced by coughing and sneezing.
- The new strain spreads rapidly because younger people have no immunity to it.
- Some older people did have some immunity, probably because they had 'memory cells' from the 1918 strain.

Every year before the seasonal flu **epidemic**, vulnerable and older people get their 'flu jab', that is, an influenza vaccination. The flu jab is different every year. It is made from the common flu virus strains that have been around during the previous season. It takes several months to identify, culture and produce vaccine for a new strain. Swine flu was identified in April 2009 but no vaccine was available until October 2009. Tamiflu and other such treatments do not cure flu but reduce the replication of the virus in the body. Taking a painkiller does not kill the pathogen but it relieves some of the symptoms, such as headaches.

Test yourself

29 Why can a person catch flu more than once?

30 Suggest why older people were immune to the 2009 strain of swine flu.

31 How does the 'flu jab' protect a person from flu virus? (Go back to Section 1.8 to answer this.)

B1 1.10 Working safely with microorganisms in school laboratories

Learning outcomes

- Know how to investigate safely the action of disinfectants and antibiotics.
- Explain how sterile culture media and transfer methods are achieved.

In school laboratories, bacteria are grown in Petri dishes on a culture medium solidified in agar gel. The culture medium provides the bacteria with the nutrients (energy source, minerals, supplementary proteins and vitamins) they need for growth. It is possible to work safely with microorganisms in school laboratories but you must observe sensible precautions when handling cultures. These are sometimes called **aseptic techniques**.

The most important precautions to take when working with bacteria are:

- Avoid contaminating a culture with 'wild' bacteria or human pathogens.

Aseptic techniques are precautions that microbiologists take when handling microorganisms to protect themselves from harmful bacteria. The techniques also keep other bacteria out of the sample being investigated.

- Work at a temperature too low to grow pathogenic bacteria. In school laboratories, the incubation temperature should be kept at 25°C. Industrial laboratories often use higher temperatures to speed up the growth of their bacterial cultures but they follow strict hygiene regulations.
- Avoid all hand-to-mouth operations, such as biting fingernails, sucking pens and licking labels.
- Prevent direct contact with any bacteria being cultured. This means that Petri dish lids should be firmly sealed with tape and not opened.

Preventing contamination of your culture

Here are some simple guidelines to prevent contamination:

- Sterilise Petri dishes and nutrient agar in an **autoclave**.
- Wash your hands and put on a clean lab coat to prevent contamination from your clothes.
- Close doors and windows to reduce air movement.
- Clear the bench where you are working and wipe it over with alcohol, which quickly destroys any bacteria.
- Lay out a clean paper hand-towel and collect all your equipment on this, including the closed agar plates, culture, inoculating loop and Bunsen burner.

An **autoclave** is an industrial steriliser. The high temperature kills microorganisms.

Making a streak plate

HOW SCIENCE WORKS
PRACTICAL SKILLS

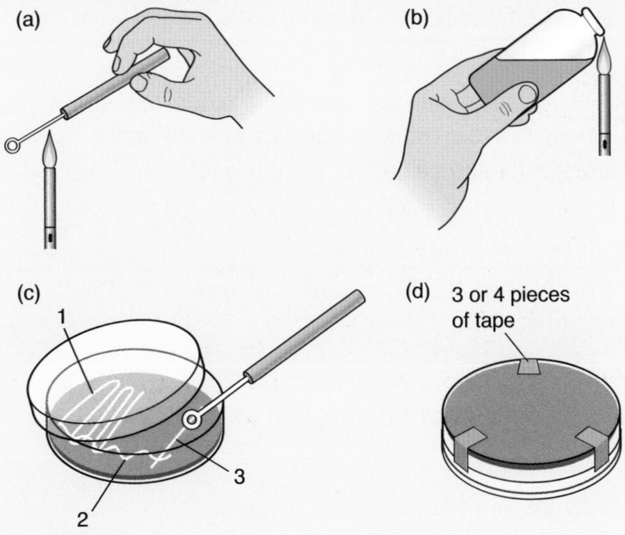

(a)

(b)

(c) 1 2 3

(d) 3 or 4 pieces of tape

Figure 1.22 The procedure for culturing bacteria in the laboratory.

Microbiologists make streak plates to identify the types of bacteria present. An inoculating loop is used to spread the bacteria over the surface of the agar. By the third or fourth streak, only a few bacterial cells are transferred. Circular colonies grow around single bacteria and this makes it easy to identify the different bacteria present.

The sequence of diagrams in Figure 1.22 shows you how to set up a streak plate from a bacterial culture.

This section also explains how to handle culture bottles to prevent contamination. **Read the whole section before starting any practical work.**

SAFETY: Wear eye protection when heating the wire loop and bottle mouth.

- Label the Petri dish on the underside — use as little writing as possible so that you can see any growths easily.
- Light the Bunsen burner that you will use to sterilise the inoculating loop. Start heating at the handle end of the wire and move towards the loop end. Heat all parts of the wire until they are red hot (Figure 1.22a).
- Be patient: allow the loop to cool in the air below the flame before dipping it in the culture solution.
- Now, holding the loop handle with your thumb and first finger, open the lid of the culture bottle using your little finger. Draw the mouth of the open bottle through the flame (Figure 1.22b). Do not put the lid on the bench. Hold it between your little finger and the palm of your hand. Dip the loop in the culture, reflame the open bottle mouth and then replace the lid. In this way you should protect the culture bottle from contamination.
- Make streak 1 on one side of the agar plate without completely removing the lid of the plate. The streak should be made on the surface (not dug into the

agar gel) as the bacteria are respiring aerobically (Figure 1.22c).

- Resterilise the loop, allow it to cool and then draw it through the tail end of the first streak. This will make a more dilute second streak.
- Repeat this dilution technique by resterilising the loop and drawing it through the tail end of the second streak to create a third streak.
- Place the loop in sterilising solution.
- Tape the lid on the plate and place it upside down in the warming cupboard to incubate until you are ready to observe the result (Figure 1.22d).
- Place all other equipment in sterilising solution. Roll up the paper towel for incineration. Wipe the bench with alcohol or disinfectant. Wash your hands and dry them on a paper towel.
- Before examining the plate, prepare your clean area as before. Do not open the plate; colonies should be visible through the base.
- When you have finished with the plate, place it in the disposal bag and it will then be autoclaved.

If you have made your streak plate successfully, the result should look like Figure 1.23.

Tip: A hot wire cannot transfer live bacteria. This is a common mistake made by students. If you touch the edge of the agar with the loop and hear hissing, it is far too hot!

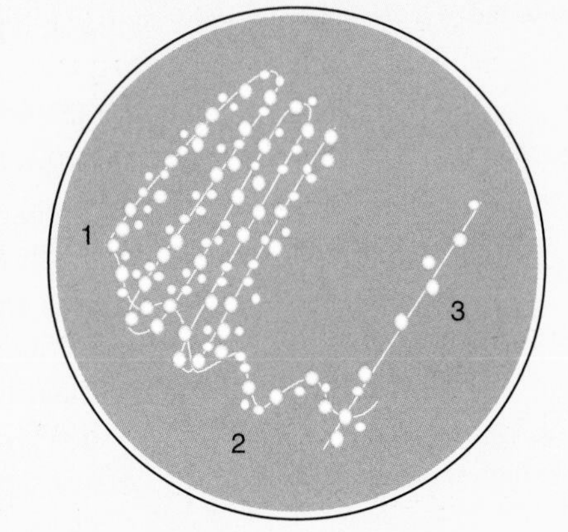

Figure 1.23 A streak plate seen from below showing thick growths of bacteria around streak 1, fewer bacteria around streak 2 and individual colonies around streak 3.

Boost your grade

If you are writing about sterile technique, don't just give instructions — explain the reason for each precaution.

Test yourself

32 When carrying out experiments with bacteria in the school laboratory, you should take precautions to avoid contamination by foreign bacteria. Give three possible sources of foreign bacteria.

33 When culturing bacteria in a school laboratory, the temperature is kept at 25°C, but when cultures are incubated in a hospital pathology laboratory, the temperature is maintained at 37°C. Explain the reasons for this difference in temperature.

34 What does agar gel contain?

35 Why should the streak be made on the surface of the agar and not dug into the agar?

36 Give two reasons why the Petri dish is sealed with tape.

Investigating the activity of different antibiotics

HOW SCIENCE WORKS ACTIVITY

This experiment was carried out by students who followed carefully all the aseptic guidelines for making a streak plate given above.

They collected the following items and placed them on a clean paper towel: Bunsen burner, fine forceps and a labelling pen. They also had two Oxoid multodiscs or combi discs ready in clean empty Petri dishes. Each multodisc is a filter paper that has been impregnated with several different antibiotics and the small circles (discs) are numbered to identify the antibiotic present using a key (see Figure 1.24).

They were given two Petri dishes that had been seeded with bacteria all over the surface and labelled with the type of bacteria.

Procedure

- Students heat-sterilised the fine forceps in the Bunsen flame and allowed them to cool below the flame.
- They picked up the multodisc with the forceps and, lifting the lid of one of the Petri dishes, placed a multodisc on the surface of the bacterial culture. The disc was pressed into position using the forceps.
- The lid was sealed on the dish with strips of tape.
- They resterilised the forceps.
- This procedure was repeated with the second Petri dish.
- Both dishes were placed in an incubator until the next lesson.
- They followed aseptic techniques for clearing up.

❶ How would you expect the antibiotic to spread from a multodisc disc into the agar?

❷ When bacteria grow on agar in a Petri dish they appear as a white film. The different antibiotics affect different bacteria in different ways (see Section 1.6). If an antibiotic prevents a bacterium from forming new cell walls during cell division, what would you expect to see around the discs?

Looking at the results

In the next lesson, the students again set up a sterile area and collected the Petri dishes but did not remove the tape or open the dishes. They used a ruler or dividers to measure the diameter of the clear area around each disc and recorded their results for each bacterium and each antibiotic in Table 1.5.

Table 1.5 Results

	Diameter of clear area (mm)				
	Disc 1	Disc 2	Disc 3	Disc 4	Disc 5
Escherichia coli (K12)	12	23	17	15	9
Bacillus subtilis	28	13	22	7	11

When clearing up they placed the used Petri dishes in the bag for autoclaving, cleared and disinfected the area and washed their hands.

Look at the results table. A bacterium is said to be susceptible to an antibiotic if the clear area has a diameter greater than 15 mm.

❸ Were all antibiotics equally successful for each bacterium?

❹ Give the number of the antibiotic that could be used to treat both types of bacterium.

❺ Give the number of the antibiotic to which each of the bacteria is most sensitive.

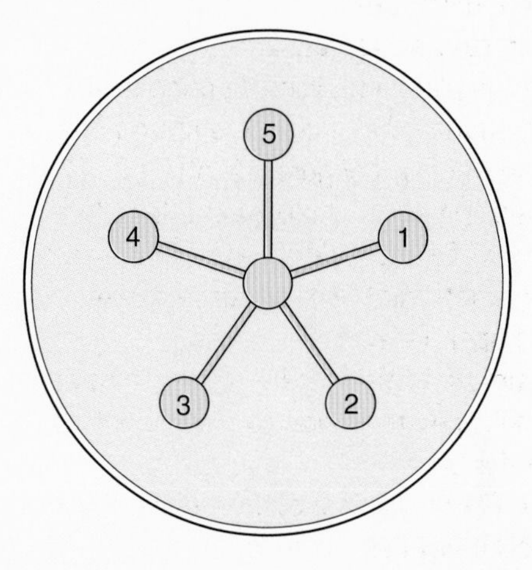

Figure 1.24 A multodisc.

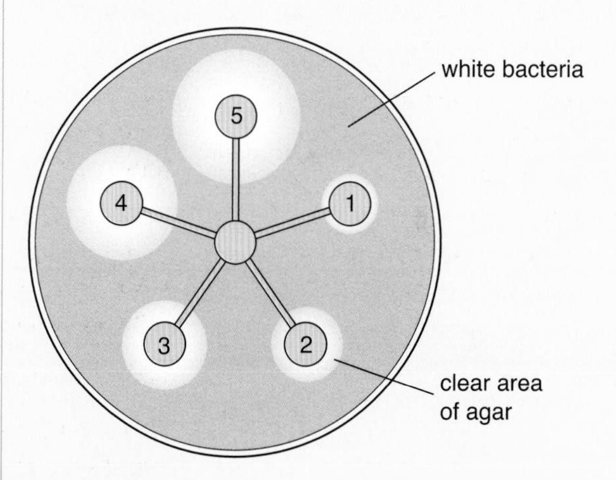

Figure 1.25 Example of the plate after incubation. Which is the most effective antibiotic?

Investigating the action of different disinfectants

The above procedure could be modified to compare the effectiveness of different common disinfectants or different dilutions of one disinfectant. You could either soak filter paper discs in the disinfectant or use a sterile pipette to add drops of the disinfectant to the Petri dishes seeded with bacteria.

❻ How would you evaluate your results?

Homework questions

1 The bar of chocolate eaten at break contains 2205 joules. Assume you weigh 60 kg. Use Table 1.2 to calculate for how long must you walk quickly to use up this extra energy. Give a reason why your answer may not be exact.

2 Vaccination programmes have been successful in preventing the spread of infectious diseases in the UK. The conditions in a developing country are very different. Use the bullet points on page 17 for ideas to explain why it is difficult to prevent the spread of an infectious disease in a developing country, such as Uganda.

3 The development of antibiotics improved the survival rate of many patients.

- **a** How do antibiotics work?

Antibiotic-resistant strains of bacteria are causing problems in most hospitals.

- **b** Why has there been a large increase in the number of antibiotic-resistant strains of bacteria?
- **c** What is the best way of reducing transmission from person to person?

4 A person eats some food contaminated with *Salmonella* (food poisoning) bacteria.

- **a** Which part of the body will these bacteria infect?
- **b** The white blood cells detect the pathogen and start to form antibodies. Draw the shape of the white blood cell that produces antibodies.
- **c** The antibodies cause the bacteria to clump together. Show, using labelled diagrams, how a second type of white cell destroys the bacteria.

5 James and John both have high cholesterol levels. They are about the same height and an ideal weight, and both play in the local football team. James eats wisely and is a vegetarian but John enjoys large steaks and chips. Explain the possible reasons for the high cholesterol levels for each of them and suggest what should be done to lower them.

6 The swine flu epidemic in the UK started in April 2009 but a vaccine was not available for several months.

- **a** What is a vaccine?
- **b** How is a vaccine made?
- **c** Why is it necessary to have a new 'flu' vaccine each year?

7 A student wrote the following: 'Bacteria and viruses are similar bugs that make you ill.'

Copy and complete the following table to compare bacteria and viruses.

	Bacteria	Viruses
Size		
Presence of nucleus		
Presence of nucleic acid		
Cytoplasm		
Outer covering		
Enzymes for nutrition		
Site of damage in human body		
Method of treatment for infection		

8 Kyle was a fussy eater. He was small and skinny for his age and got tired easily. He ate only toast and jam, crisps and chips washed down with cola.

- **a** Do you think this is an adequate diet for a 13 year old?
- **b** Kyle had a permanent cold. What two missing food groups could explain this?
- **c** What missing mineral could explain why he tired easily? What is the role of this mineral in the body?

Kyle felt left out and desperately wanted to be in a sports team. The PE master was encouraging and said Kyle needed to be on a 'sports diet'. The master recommended the following: eggs on toast with a glass of orange juice for breakfast; meat, green vegetables and baked potato for lunch; fish, salad and pasta for an evening meal. Kyle was allowed snacks such as apples and cheese, muesli bars, milk and bananas. He stuck to diet changes and a year later he was selected for the second eleven team.

- **d** Explain how each of the items in his new diet resulted in such an improvement in his physique.

Exam corner

a **The child vaccination programme aims to vaccinate 95% of children. This reduces the number of cases of common childhood diseases. A child who was not vaccinated caught measles. He was very unwell for a few days and then steadily got better. Why did he feel ill for a few days and then why did he get better?** *3 marks*

b **The child was told that he would not get measles again. Why can't you catch measles twice?** *3 marks*

c **Some of the children's teachers caught 'flu' for a second time. Explain how you can catch some virus infections more than once.** *3 marks*

Student A

a He felt ill because the measles virus in the cells was replicating fast and destroying the cells. ✓ The body detected the virus and made antibodies ✓ that only work on measles virus. Detecting and making the antibodies takes a few days. When there were enough antibodies he got better. ✓

Examiner comment This is a good answer explaining clearly how a virus damages cells. Although the student refers to 'the body', he/she does explain that antibodies are made and these are specific to the measles virus. A reference to white blood cells would have strengthened the answer. The time lag is explained well. This answer scores all 3 marks.

Student B

a He felt ill because he had a virus. The body made antibiotics ✗ which killed the bacteria and then he felt better.

Examiner comment This student makes three common mistakes. First, he/she confuses the words bacteria and virus — they are not the same. Second, he/she has not read and answered the question set. There is no explanation as to why the child felt unwell or why he subsequently got better. The wording of the question and 3 marks for the answer indicate that an explanation is required. Finally, words that sound similar need to be practised. Student B has confused antibodies and antibiotics. Student B fails to score.

Student A

b At the same time as antibodies are made, so are memory cells. ✓ These cells remember the shape of the virus and if the same virus gets into the body again, the memory cells switch on the system ✓ that makes the special antibodies. So a lot of antibodies are made quickly. ✓

Examiner comment Student A has remembered the production of memory cells for 1 mark and that they remember the shape of the virus, which gained a mark in Q1 so the mark is not given again. However, a mark is awarded for switching on the antibody production. Another mark is given for the role of memory cells in the rapid production of large numbers of antibodies. Again, there is no mention of white blood cells, or the fact that the antibodies destroy the virus before there are enough to make the child ill.

Student B

b Antibodies are made very quickly a second time ✓ and destroy the virus before the person gets ill. ✓

Examiner comment This is a much better answer although it does not use sufficiently technical language. The student has remembered that antibodies are produced quickly the second time and here the correct word is used. The second mark is awarded for the antibodies destroying the virus.

Student A

c Flu viruses frequently mutate. ✓ The mutation causes a change in the surface protein ✓ so the teacher became ill again.

Examiner comment Student A has correctly referred to mutations and could have included the word 'random'. The result of the mutation in changing the surface protein would have been stronger if the answer had referred to shape. However it is just sufficient for 1 mark. The answer is incomplete as there is no explanation of why previous antibodies do not function.

Student B

c Some viruses often change their genetics, which gives them a different shape. ✓ The old antibodies don't know the new shape and can't work on it so the person gets ill. ✓

Examiner comment Student B understands the idea of mutation but unfortunately does not use the word — a weak mark is allowed. The reason why the previous antibodies can't function is well stated.

Chapter 2
Nerves and hormones

Setting the scene

The participants in this white water rafting activity are receiving and responding to many stimuli. How many stimuli can you identify?

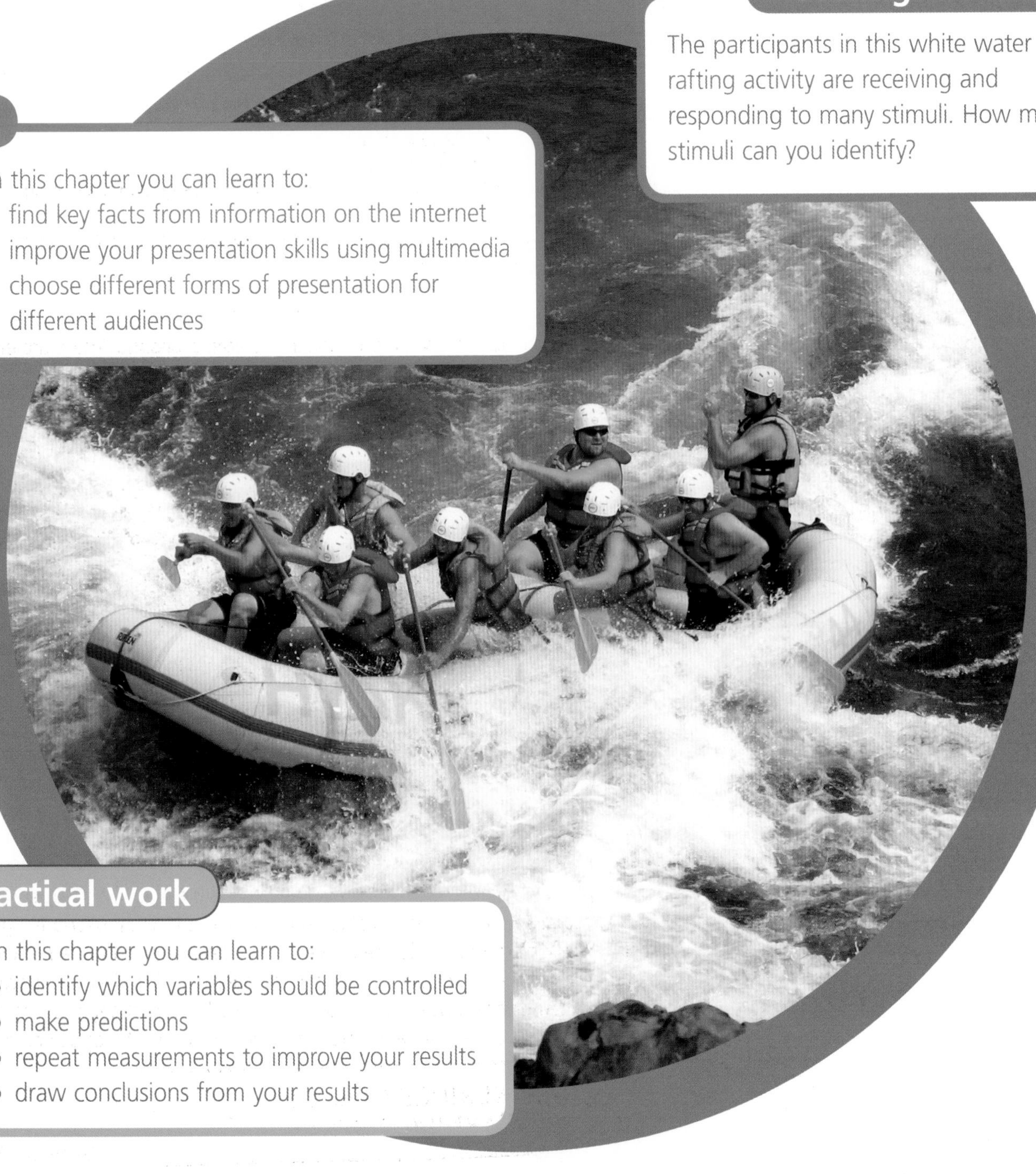

ICT

In this chapter you can learn to:

- find key facts from information on the internet
- improve your presentation skills using multimedia
- choose different forms of presentation for different audiences

Practical work

In this chapter you can learn to:

- identify which variables should be controlled
- make predictions
- repeat measurements to improve your results
- draw conclusions from your results

B1 2.1 The nervous system

Learning outcomes

- Know how the different parts of the nervous system work together to coordinate our responses to external stimuli.
- Know the names of receptors for different stimuli.

Your body is constantly responding to changes taking place outside and inside your body. These changes or inputs are known as **stimuli**. If someone throws a ball to you, you respond quickly by moving your hands to catch it. If a teacher is about to ask you a question in class, you may feel nervous.

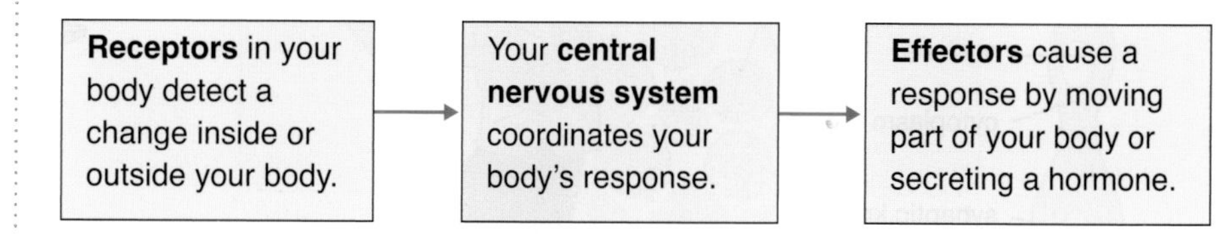

Figure 2.1 This flow diagram shows how the nervous system enables the body to respond to changes.

Worked example

Coordination of a voluntary response to a stimulus

- **Stimulus** — a friend calls your name.
- **Receptors** in your ears detect the sound.
- An **impulse** passes from your ears to your **central nervous system** (CNS) along a sensory neurone.
- The **CNS** registers the stimulus and coordinates the response by sending an impulse to your muscles along a motor neurone.
- **Effector** muscles contract so that you turn and face your friend.

Receptors are cells or organs that can detect changes inside or outside the body.

The **central nervous system** (CNS) is the spinal cord and the brain.

Effectors are organs in the body that cause a response. They can be muscles (which respond by contracting) or glands (which respond by secreting hormones).

Sense organs are organs that contain receptor cells.

A **stimulus** (plural: **stimuli**) is a change detected by receptor cells.

An **impulse** is a rapid change in the ions around a neurone which travels from one end of a neurone to the synapse.

Some receptors in your body are grouped in the **sense organs**. There are eight different types of receptors in your body. These are shown in Table 2.1.

Table 2.1 The range of receptors found in the body and the stimuli they detect.

Stimulus	Receptor	Sense organ
Light	Light receptors	Eye
Sound	Sound receptors	Ear
Movement	Position receptors Touch receptors Pressure receptors	Ear Skin Skin
Chemicals	Chemical receptors for taste and smell	Tongue Nose
Change in temperature	Temperature receptors	Skin Blood vessels
Pressure/temperature	Pain receptors	Skin

A special receptor organ

Light receptor cells at the back of the eye respond to changes in light intensity or colour. Those in the centre detect colour and those at the side respond to light intensity. The light receptor cells are typical animal cells — they have a nucleus and cytoplasm and are surrounded by a cell membrane — but they also have some extra membranes inside, which have a chemical that is changed by light. The chemical change starts a nerve impulse, which passes to the optic nerve across the synapse.

Voluntary actions are responses that are coordinated by the brain.

Reflex actions are automatic, rapid responses to danger.

Sensory neurones carry impulses from a receptor or sense organ to the spinal cord or brain.

Relay neurones pass the impulses from sensory neurones to motor neurones.

Motor neurones carry impulses to effectors such as muscles or glands.

In a **spinal reflex** the impulse passes from the sensory neurone to the motor neurone via the relay neurone. It does not travel via the brain — it is automatic.

How fast can the nervous system respond to stimuli?

Nerve impulses can travel along your neurones at about 100 metres per second. That's the speed you would need to run to finish a 100 m race in 1 second! So, all responses are fast but for many responses you need to add the time it takes for the brain to think about the information and coordinate a response. Responses that are coordinated by the brain are called **voluntary actions**. An example of a voluntary action would be you responding to a question from your friend.

If you touch a very hot object, you pull your hand away before thinking about it. This type of rapid, automatic response to danger is called a **reflex action**. **Motor neurones** may also connect to glands — for example, if you are out in the dark and there is a sudden scary noise, the motor neurone from the CNS connects to the adrenal gland, which quickly secretes adrenaline to prepare the body to run away — this is a reflex action.

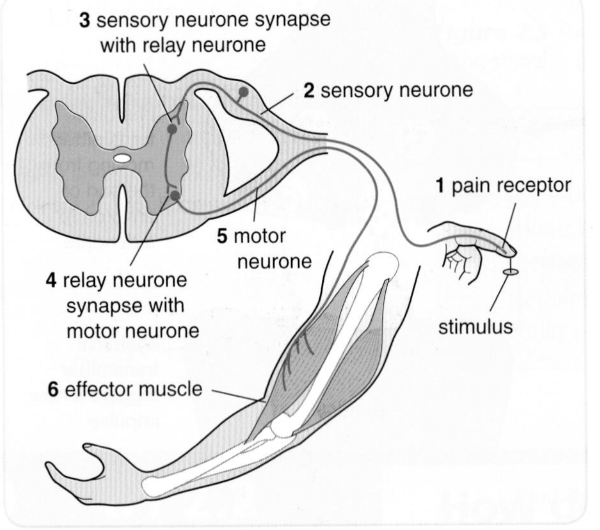

Figure 2.7 The neurones involved in a **spinal reflex**.

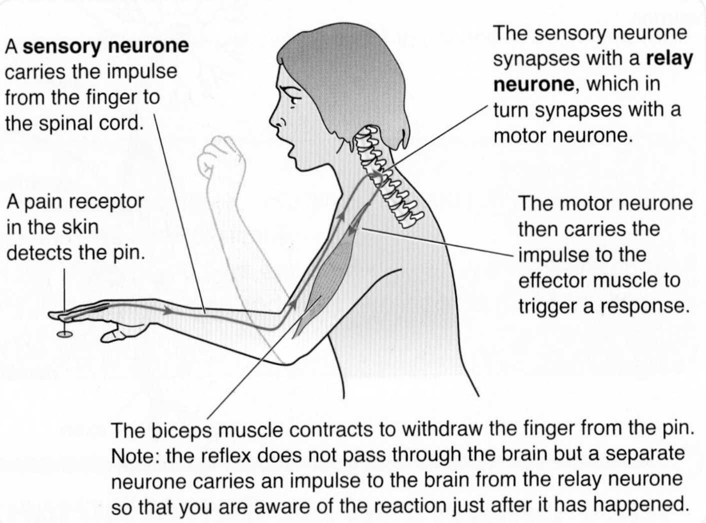

A **sensory neurone** carries the impulse from the finger to the spinal cord.

The sensory neurone synapses with a **relay neurone**, which in turn synapses with a motor neurone.

A pain receptor in the skin detects the pin.

The motor neurone then carries the impulse to the effector muscle to trigger a response.

The biceps muscle contracts to withdraw the finger from the pin. Note: the reflex does not pass through the brain but a separate neurone carries an impulse to the brain from the relay neurone so that you are aware of the reaction just after it has happened.

Figure 2.8 A spinal reflex. Note the route.

Test yourself

6 Which of the responses listed below are voluntary actions and which are reflex actions?
- Blinking when someone kicks dust at you.
- Sneezing when pepper goes up your nose.
- Picking up a pen.
- Pulling your hand away when you touch a hot iron.
- Turning around when someone calls your name.
- Taking a CD off a shelf.

7 Choose one of the reflex actions from question 6. Write out a sequence for the action starting with:

A __________ receptor detects ___________________.

Multiple sclerosis

HOW SCIENCE WORKS
ICT

Multiple sclerosis (MS) is a disease that damages the central nervous system (brain and spinal cord). Its symptoms are varied and can include blurred or double vision, slurred speech, poor coordination, muscle weakness and paralysis. At present there is no cure for multiple sclerosis but a lot of research is being carried out to develop treatments for the disease.

The nerves in the central nervous system are coated with a chemical called myelin. This insulates the nerve in a similar way to the plastic coating on electrical wire. When a person suffers from MS, patches of myelin become damaged, exposing the nerve itself.

Use these websites to help answer the following questions, particularly question 5:

www.mssociety.org.uk — this website has a section about childhood MS

www.msrc.co.uk — this website is a bit more advanced but has a useful YouTube video and a link on paediatric MS research

www.mstrust.org.uk — the 'Any Questions' section has information on exercise

1. What are the symptoms of MS?
2. From what you have learnt about the nervous system, suggest how a damaged myelin coating can cause the symptoms of MS.
3. MS causes damage to nerves only in the brain and spinal cord. However, a sufferer may have poor coordination of their legs. Why does this happen if the nerves in the legs are not actually damaged?
4. Muscle weakness and fatigue in MS are caused by poor transmission of nerve impulses. Suggest why it is now considered important to exercise the muscles.
5. Use websites to prepare a list of six 'frequently asked questions', with answers, that could provide useful information for young people who have recently been diagnosed with MS. You could design your own web page for children with MS or just make a poster. Think about the information that would really help.

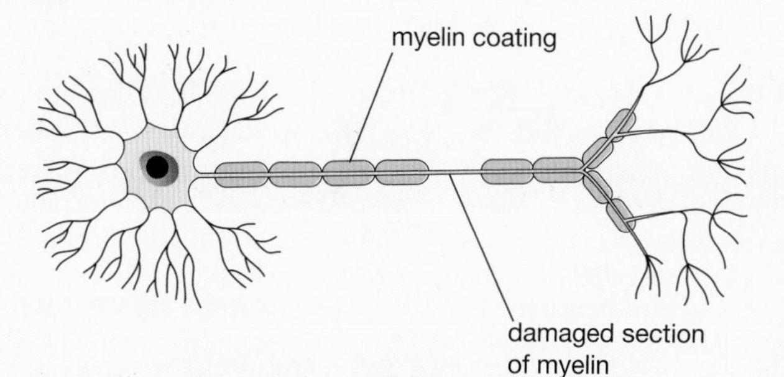

Figure 2.9 A nerve with a damaged myelin coating in a person suffering from MS. Why is this neurone not working properly?

Reaction times

HOW SCIENCE WORKS
PRACTICAL SKILLS

How could you measure the reaction times of your classmates using just a 30 cm ruler?

Does practice improve reaction times in your experiment?

Share your results with the rest of your class.

Can drugs such as caffeine affect your results? How could you find out?

B1 2.3 Control in the human body

Learning outcomes

- Know that internal conditions are controlled within a narrow range.
- Know the basic sequence of hormone control.
- Understand why body temperature must be closely controlled.
- Understand that chemicals called hormones coordinate many processes in the body.

Hormones are chemicals that are released from glands into the blood and then transported to their target organs.

Glands are the parts of the body that release (secrete) hormones.

Each hormone acts on a specific **target organ**.

Enzymes control many of the chemical reactions that take place in the body. Enzymes work best under certain conditions and body temperature is controlled within a narrow range for these enzymes. Water content, ion content and blood sugar levels are also internal conditions that are closely regulated. **Hormones** control much of this regulation.

What are hormones?

When asked about hormones, people often mention teenage mood swings and puberty. Hormones trigger both these changes. This section will help you to explain how hormones control some of the internal conditions of your body. Organs that release hormones are called **glands**. Hormones act on specific organs called **target organs**.

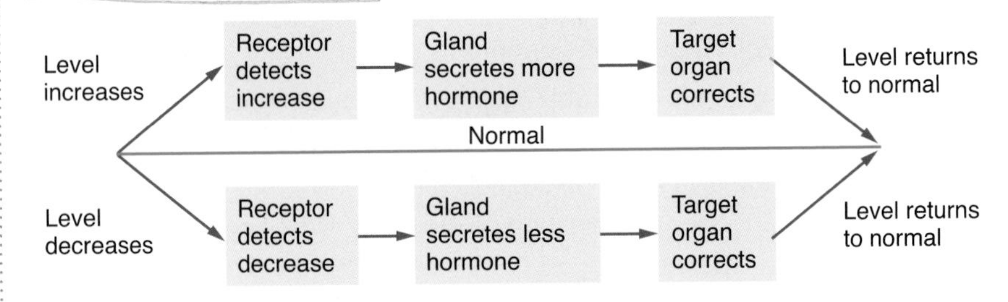

Figure 2.10 This basic sequence applies to all hormone control.

How do hormones control conditions inside your body?

Hormones are involved in most of the control mechanisms in the examples in Table 2.2.

Table 2.2

Internal condition	How it changes	Why it must be controlled	How it is controlled
Water content of the blood	Water is gained through food and drink Water is lost from the lungs when exhaling, when sweating and in urine	Substances are transported in the blood All reactions take place in solution Water is used for cooling the body when sweating Waste substances are excreted in urine	Hormones regulate water loss and conservation by the kidneys, which are the target organ
Ions in blood	Gained in food and drink	Lost in sweating Lost in urine	Hormones regulate ion loss via the kidneys
Temperature	Temperature rises with muscular exercise Heat is lost from the skin surface, lowering body temperature	Enzymes work best within a narrow range of temperatures	By changing the blood flow through capillaries at the skin surface Hormones increase metabolic rate in cells Shivering Sweating
Sugar level in the blood	Gained in food and drink Used during muscular exercise Used to maintain body temperature	To provide cells with a constant supply of glucose for making cellular energy The brain uses sugar for energy	Hormones control storage of excess sugar (target organ is the liver) and release of stored sugar from the liver and muscles

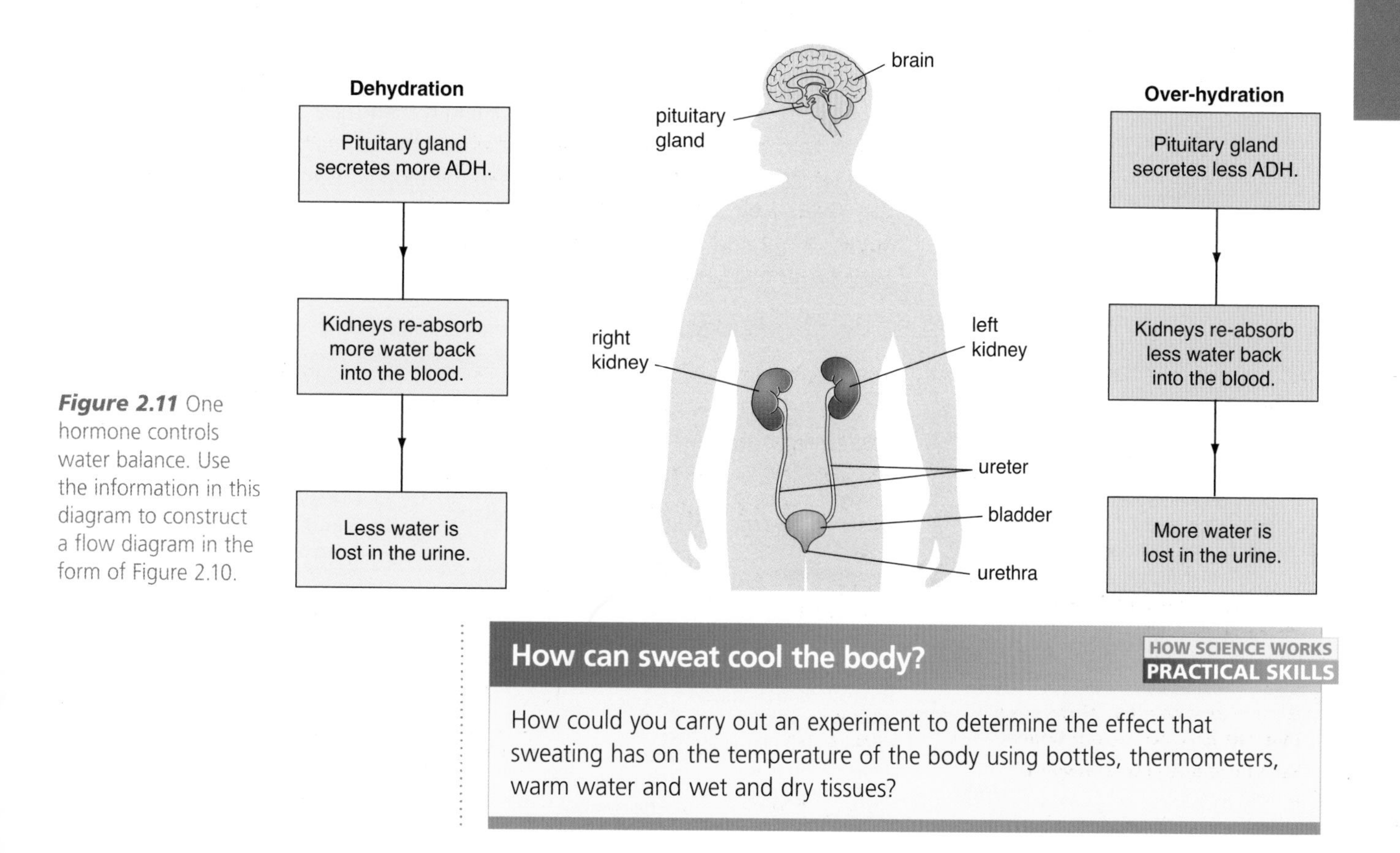

Figure 2.11 One hormone controls water balance. Use the information in this diagram to construct a flow diagram in the form of Figure 2.10.

How can sweat cool the body?

HOW SCIENCE WORKS
PRACTICAL SKILLS

How could you carry out an experiment to determine the effect that sweating has on the temperature of the body using bottles, thermometers, warm water and wet and dry tissues?

B1 2.4

Learning outcomes

- Know how hormones can be used to control a woman's fertility.
- Understand how hormones control the menstrual cycle.

How do several hormones work together to control the menstrual cycle in women?

The menstrual cycle is controlled by changes in hormone levels. A hormone reaches a level that stimulates the release of another hormone (rather like one domino falling and knocking the next over). This means that hormones respond to both normal cycles and also to cycles that result in pregnancy.

The cycle starts with the release of follicle-stimulating hormone (FSH) from the pituitary gland, which is the master control gland in the brain.

A woman's menstrual cycle lasts about 28 days. At the start of a cycle the lining of the uterus

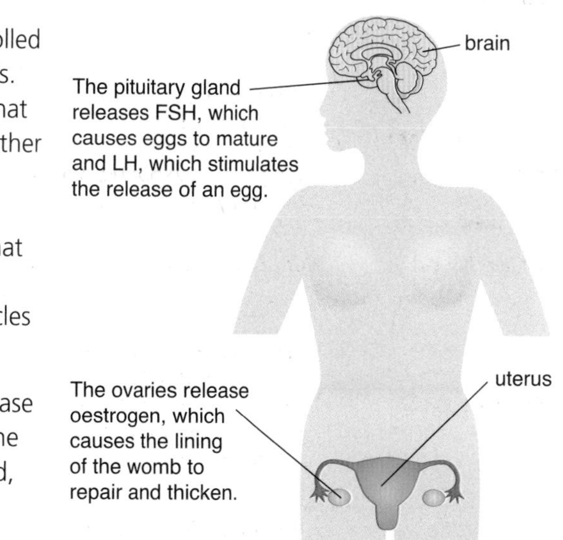

Figure 2.12 Hormones produced by the pituitary and ovaries control the uterus. Examine carefully the sites of production of the hormones and the target organs.

(womb) thickens. An egg matures and is released from the ovary. If this egg is fertilised by a sperm cell it may implant in the lining of the womb. If the egg does not implant in the womb much of the lining is then shed during the woman's period. Hormones released by the ovaries and the pituitary gland control all these changes. The levels of these hormones in the bloodstream are different before and after ovulation. This is a delicate control system and it is easy to upset the menstrual cycle by over-exercise or stress. It is necessary to know about and understand the female hormones for birth control and fertility treatment.

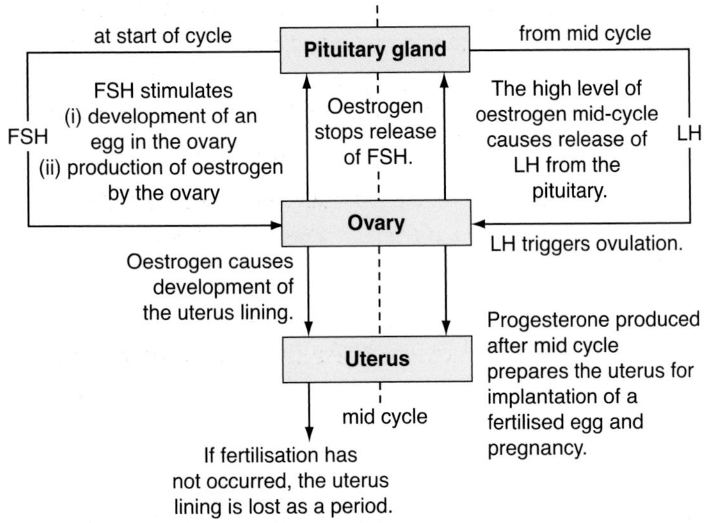

at start of cycle

Pituitary gland

from mid cycle

FSH

FSH stimulates
(i) development of an egg in the ovary
(ii) production of oestrogen by the ovary

Oestrogen stops release of FSH.

The high level of oestrogen mid-cycle causes release of LH from the pituitary.

LH

Ovary

LH triggers ovulation.

Oestrogen causes development of the uterus lining.

Progesterone produced after mid cycle prepares the uterus for implantation of a fertilised egg and pregnancy.

Uterus

mid cycle

If fertilisation has not occurred, the uterus lining is lost as a period.

Figure 2.13 Hormones control the menstrual cycle.

Table 2.3 The main hormones involved in control of the menstrual cycle.

Hormone	Released by	Target organ and effect
Follicle-stimulating hormone (FSH)	Pituitary gland	Ovary • Causes an egg to mature in the ovary • Stimulates ovaries to produce oestrogen
Oestrogen	Ovaries	Uterus • Causes lining to thicken in first half of the cycle Pituitary • High oestrogen level switches off release of FSH and switches on release of LH
Luteinising hormone (LH)	Pituitary gland	Ovary • Stimulates ovulation (release of the egg from the ovary)
Progesterone (produced if egg is fertilised)	Ovaries	Uterus • Maintains thick uterus lining if the egg is fertilised • High levels of progesterone in pregnancy stop the cycle

Test yourself

8 Fill in the blanks to help learn the sequence of hormones:

FSH released from the ______ goes to the target organ, which is the ______, and stimulates the release of ______.

High levels of ______ travel to the ______ and switch on release of ______ and switch off release of ______.

______ travels to the target organ (the ovaries) and causes ovulation.

If the egg is fertilised, ______ is produced from the ovaries and this hormone maintains pregnancy.

How can hormones be used to control fertility?

Increasing the chance of pregnancy by fertility treatment

Sometimes a woman's pituitary gland does not produce enough FSH at the beginning of a cycle. An artificial 'fertility drug', which acts like FSH and LH, can be given to stimulate the production and maturation of the egg. This is given by injections at the start of the cycle. It increases the chance of producing more than one egg and carries the risk of twins or multiple embryos. This hormone is also used in *in vitro* fertilisation (IVF) treatment.

Reducing the chances of pregnancy — oral contraception

A high level of oestrogen stops the production of FSH at the mid-cycle. The oral contraceptive pill contains a combination of oestrogen and progesterone which stops the production of FSH. Without FSH the eggs will not mature and with no egg there will be no pregnancy.

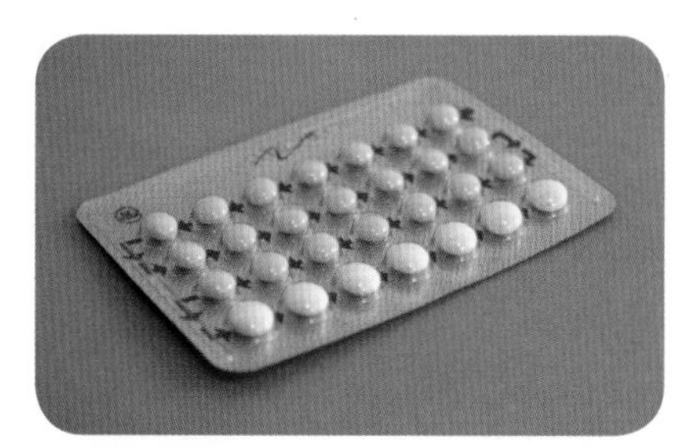

Figure 2.14 A combined pill has 21 normal pills and 7 'dummy' pills.

The original birth-control pills (known as 'the Pill') were high doses of synthetic oestrogen only. These caused side effects such as blood clots, heart attacks and strokes in some women. Research led to the development of pills with reduced levels of oestrogen that, when combined with progesterone, produced a safe contraceptive if taken correctly. The use of contraception has enabled women to pursue a career after marriage and have children when they are older.

Mini-pills contain progesterone only but are not as reliable as the combined pill. They cause fewer side effects and are ideal for older mothers for whom another child would not be a problem.

The important fact to remember is that oral contraception does not give any protection against sexually transmitted infections. These can be very serious and are easily passed from one person to another.

Using hormones to control fertility

HOW SCIENCE WORKS ACTIVITY

The hormones involved in the menstrual cycle can be used to help a woman become pregnant or to prevent this happening (contraception). Hormones in the contraceptive pill reduce fertility and prevent a woman from becoming pregnant.

Laura and David have just got married. They plan to have children in about 3 years' time but until then they want to have sex without Laura getting pregnant. Laura talked to her doctor about possible methods of contraception. The doctor suggested that Laura should go on the pill and explained that the pill contains a hormone that would prevent an egg maturing in the ovaries each month.

1. Why do you think Laura's doctor suggested the pill as a method of contraception? Discuss this in a pair and write a list of reasons for using the pill.
2. What possible disadvantages are there in using the pill as a method of contraception? Use the following website to help you answer this: www.fpa.org.uk. Click on the following links: 'Contraception and sexual health guide', 'Contraception', 'The combined pill', 'Are there any risks?'
3. The pill contains the hormone oestrogen. Use Figure 2.13 and Table 2.3 to explain how oestrogen can prevent an egg from maturing.

After 3 years Laura and David decided to start a family so Laura stopped taking the pill. They had sex regularly, but after a year Laura was still not pregnant. Her doctor referred them to a fertility clinic. After some tests the doctor found that the infertility was due to Laura's eggs not maturing properly in the ovaries. This was caused by a low concentration of one of her hormones. She was prescribed a fertility drug that contained a synthetic form of the hormone. The hormone caused her eggs to mature properly in the ovaries. After 6 months Laura became pregnant.

4. Look at Table 2.3. What hormone do you think the fertility drug contained? Explain your answer.
5. a Some people do not agree with the use of synthetic hormones to treat infertility. Their reasons may be based on:

   A evidence
   B hearsay
   C prejudice
   D personal opinion

Look carefully at the reasons given by the following four people for not agreeing with the use of synthetic hormones to treat infertility. In each case, decide which of A, B, C or D their reason is based on.

(i) Kelly says 'It's wrong to interfere with nature'.
(ii) Her husband Ali says 'The use of synthetic hormones could create problems in Kelly's hormonal system'.
(iii) Becky says 'It's against my religious beliefs'.
(iv) Her friend Lisa says 'The synthetic hormones are drugs and all drugs are bad for you'.

b What is your opinion on the use of synthetic hormones to treat infertility?

In some cases of infertility, taking this hormone does not help a couple have children. Another treatment is called IVF (*in vitro* fertilisation). This involves the woman taking an additional hormone to stimulate egg production. Eggs are then removed from her ovaries with a pipette and fertilised with the man's sperm in a laboratory. The eggs are grown until they form a tiny ball of cells. Two of the fertilised eggs are then inserted into the woman's womb in the hope that at least one of them will implant and develop into a baby:

www.cks.nhs.uk/patient_information_leaflet/ivf

6 What causes of infertility do you think IVF can be used to treat?

7 Suggest two potential problems related to IVF.

B1 2.5 Plants have hormones too

Learning outcomes

- Know that plants are sensitive to some environmental stimuli.
- Understand how plants respond to these stimuli.
- Explain how hormones coordinate growth in plants.
- Evaluate the use of plant hormones in horticulture and agriculture as weedkillers and rooting powders.

Test yourself

9 a Where is the plant hormone auxin produced?
b What is the effect of light on this hormone?
c What is the name given to a response to light?

Plants are sensitive to light, moisture and gravity. You will have noted that the shoots of a pot plant on a windowsill bend to face towards the outside and the pot has to be turned to keep them growing straight. If a pot is knocked onto its side the shoots turn upwards. These changes are a result of different concentrations of the growth hormone called auxin.

How does auxin control the response of a shoot to light?

Auxin is a plant hormone. It is produced in the growing tip and diffuses downwards to the sensitive cells behind the tip (this can be compared to the target organ for animal hormones). With increased auxin concentration these cells elongate more, i.e. the stem grows more.

With light from directly overhead, the auxin diffuses downward evenly all around the stem and the stem grows straight up. If the plant shoot tip is illuminated from one side the shoot will curve towards the light (Figure 2.15). This is because the light reaching the side of the shoot tip causes the auxin to move across the stem to the dark

Figure 2.15 Shoot response to light from one side. Notice how it is growing towards the light.

side. The increased concentration causes greater cell elongation on the dark side so the shoot bends towards the light. This response of the shoot tip bending towards the light stimulus is called positive phototropism.

Figure 2.16 If an entire young plant is laid horizontally, this is what is seen. Which way has the shoot bent? Is it towards or away from the gravitational pull? Is this positive or negative gravitropism?

How does auxin control the response of a root to gravity?

The auxin passes down the stem to the root where its concentration is low. Higher concentrations slow down root cell growth. If a root is growing vertically all sides receive equal concentrations and the root grows straight down. If the root is horizontal, the concentration of auxin is higher on the underside (there is experimental evidence for the concentration difference but the exact mechanism is not known). The high concentration inhibits cell elongation in the sensitive region, so the underside grows more slowly and the root turns down. The response of the root tip bending in the direction of the gravitational force is termed positive gravitropism (or positive geotropism).

Horizontal growth

HOW SCIENCE WORKS
PRACTICAL SKILLS

How might you 'confuse' a plant into growing horizontally? Plan and carry out your experiment.

A puzzling observation

In a germination experiment seeds are placed on damp tissue in a Petri dish. The dish is placed on a shelf. Two days later, all the roots are pointing upwards and do not show the expected positive gravitropism! The lid of the dish is covered in condensation. What do you think the roots were growing towards? Have you worked it out? The roots were showing positive hydrotropism — growing towards the water. Normally they would be seeking water down in the soil.

What is the importance or benefit to the plant of the hormone-coordinated growth responses?

Shoots grow towards the light because they need light energy for photosynthesis.

Roots grow in the direction of the gravitational force because they need to anchor the plant into the soil to obtain water and mineral ions.

Roots grow towards water to obtain this necessary substance for supporting the plant cells, as a raw material for photosynthesis and because all reactions take place in the watery cytoplasm.

Uses of plant hormones in horticulture and agriculture

Rooting hormone used when taking cuttings

Artificial plant hormones are regularly used for taking cuttings. You may have taken cuttings to increase the numbers of a plant more quickly than using seed. Simple shoot cuttings are made as shown in Figure 2.17.

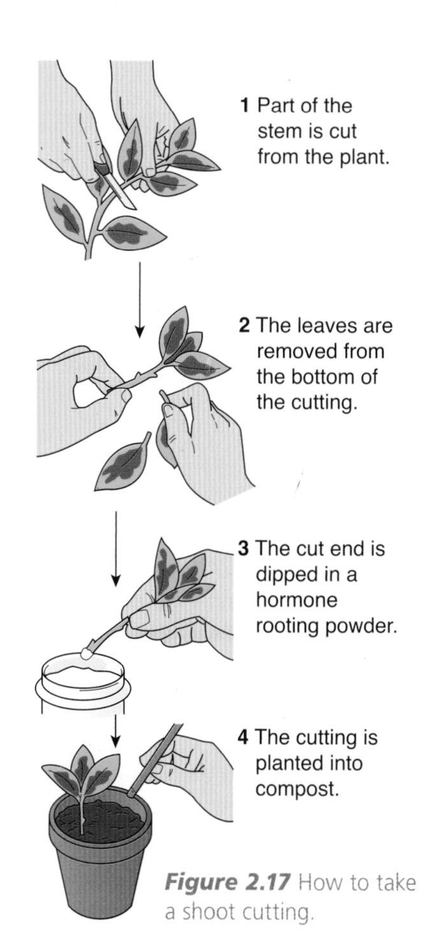

Figure 2.17 How to take a shoot cutting.

Testing hormone rooting powders

HOW SCIENCE WORKS
PRACTICAL SKILLS

Plan and carry out an experiment to compare the effectiveness of different rooting powders.

Before trying out your experiment, discuss it with your teacher, who will advise you on how to carry it out safely.

Try this technique on common houseplants such as pelargonium, dahlia and fuchsia. Dipping the cut end of the shoot in rooting powder stimulates the development of roots — you will end up with a complete plant that is identical to the parent plant.

Selective weedkiller

Synthetic auxins are used as weedkillers. They are taken into the plant and transported in the same way as natural auxins. Synthetic auxins cannot be broken down by the plant's enzymes and so reach high concentrations in the cells. This causes rapid distorted and uncoordinated growth of some parts of the plant while inhibiting the functioning of other parts, so the weed dies. Selective weedkillers are effective against broadleaved plants such as dandelions and daisies growing in a lawn or on a golf course. The large surface area of the flat leaves results in dandelions taking up a large quantity of the hormone but the spray rolls off the vertical rolled leaf of the grass, leaving it unharmed.

Figure 2.18 What difference has treatment with weedkiller made?

Test yourself

10 Why does a hormone weedkiller kill off daisies but leave the grass unharmed?

B1 2.6 How are new medical drugs developed?

Learning outcomes

- Explain how medical drugs are developed and tested.

We have seen how chemicals in the form of natural and synthetic hormones can bring about changes in the body. Now we will look at some different chemicals under the general name of drugs. Drugs are chemicals that have an effect on processes in the body. Medicines are drugs. When used properly, they are beneficial to the body. Some people use substances such as nicotine, alcohol, cannabis and heroin as recreational drugs. Some of these are more harmful than others. It is important to know exactly how a drug affects our bodies before we can assess its possible risks or benefits.

Medical drugs

Medical drugs or medicines are used to treat disease, injury and pain. Medicines are beneficial when used properly but can still be harmful if misused. Paracetamol is an effective painkiller if you take the correct dose, but taking more than this can cause liver damage. In extreme cases, it can kill. This is why it is essential that you always follow the instructions given with medicines.

Test yourself

11 Why is it important to carefully follow the instructions when you take medicines?

12 Why do you think it is important to medical science that the rainforests are preserved?

More than 7000 chemicals used to make medicines are found in plants and many of these plants are found in the rainforests. The indigenous people of the rainforest use the plants growing there to treat many illnesses. Quinine is extracted from the *Cinchona* tree, found in South America, and is used to treat malaria. A plant called rosy periwinkle, found in Madagascar, provides two of the most important chemicals currently used to treat cancer. New drugs are still being developed from natural substances found in plants.

The testing and development of modern medicines and drugs

HOW SCIENCE WORKS ACTIVITY

The process of developing and testing modern medicines and drugs illustrates the way in which research scientists work. It often begins with an idea or an explanation using existing theories as to why a particular drug or medicine might be used to treat a specific disease. This idea or explanation is called a **hypothesis**. The hypothesis can then be used to make predictions about the effects of the drug that can be thoroughly tested.

When scientists develop a new drug to treat a disease it must be thoroughly tested. The testing is carried out in a number of stages.

- Stage 1: The researcher knows how the disease affects cells and can therefore select possible chemicals that will interfere with the disease mechanism. This involves some careful chemical simulation on a computer to test the hypothesis.
- Stage 2: Drugs are tested in the laboratory. They are tested first on live cells and then sometimes on animals to find the level of toxicity. These tests are used as a model to predict how the drug may react in humans. The predictions are tested in clinical trials.
- Stage 3: Drugs that show promise in the laboratory are tested in clinical trials.
  - In Phase I of these trials, drugs are tested on a small number of healthy human volunteers, starting with low doses, to find if there are any side effects.
  - If there are no side effects the drug moves to Phase II testing. The drug is tested on about 100 humans who suffer from the disease that the drug is intended to treat.
  - In Phase III of clinical testing the drug is tested on several hundred sufferers and its effectiveness is compared with an existing drug for the disease, and also to find the optimum dose.

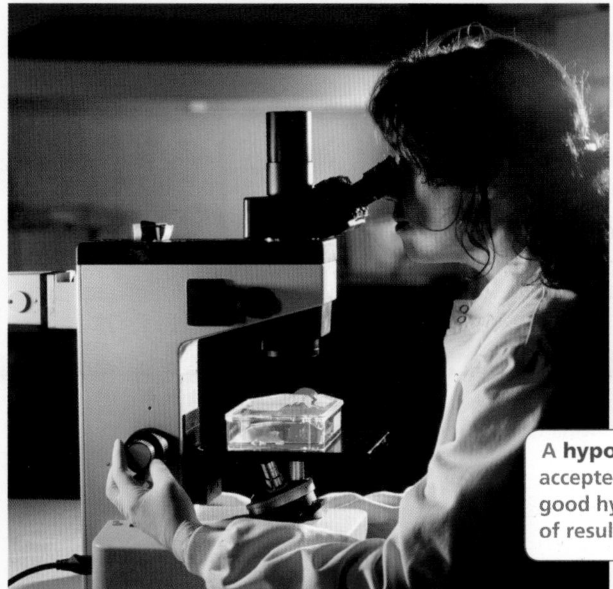

Figure 2.19 Looking at the effect of a drug on live cells.

1. What are the advantages of testing a new drug on live cells instead of on animals?
2. In Phase I of the testing process why do you think new drugs are tested on healthy people instead of people with the disease?
3. Why does testing at Phase II start with very low doses?
4. Why do you think the majority of tests take place on young males? What possible problems could arise?
5. Suggest reasons why scientists test a new drug against an existing medicine in Phase III of the process.

A **hypothesis** is an idea or explanation based on accepted scientific theories (it is a testable idea). A good hypothesis should be able to make predictions of results that can be tested.

Clinical trials in detail

HOW SCIENCE WORKS ACTIVITY

If the experiments and observations support the initial hypothesis about the drug, then clinical trials Phase II can begin with sick patients. Look below at the care taken in these trials.

These trials will use carefully selected groups to check the effects of the drug on people of different sex, age and weight.

They will also involve control groups to ensure that only the effects of the drug (the independent variable) are observed. The control group will be given a placebo (tablets that look like the drug but have no effect).

1. Suggest why placebos are used.

In a double-blind trial the patients do not know whether they are receiving the experimental drug or the placebo — both are packaged identically so even the research doctor does not know.

2. Give two reasons why the placebos appear identical to the trial drug.
3. What effect might result if the research doctor knew what the patient was taking?

All effects and symptoms are recorded and analysed at the end of the trial to evaluate the effectiveness of the new drug.

4. At the end of the trial, the same percentage of patients on the placebo as on the drug felt better. What can you conclude?

In a separate trial, during Phase III testing, one group of patients took a tried and tested medication and a separate group took a new medication. The new medication was far more expensive.

5. What do you think the doctors who prescribed the medication would be looking for in the results of the tests?

B1 2.7

Learning outcomes

- Understand the time taken to develop and test medicinal drugs.
- Know that statins lower the amount of cholesterol in the body.

Testing is vital

If after all these tests and trials the results support the initial predictions about the drug, then large-scale production can begin. Even when the drug has been licensed for use, any possible side effects are reported and carefully monitored. The development, testing process and clinical trials take about 10 years and cost millions of pounds. Only those drugs that pass all safety checks and clinical trials can be launched as a new product. In most cases, medical drugs that are used according to the instructions have no, or only limited, side effects.

A drug taken for uses for which it was not tested

Occasionally a drug that has been tested causes unexpected side effects. A drug called thalidomide was used in the 1960s as a sleeping pill. It was also effective at relieving morning sickness in pregnant women but had not been tested for this use. Unfortunately, many babies born to mothers who took the drug had underdeveloped arms and legs because it reduced the development of blood capillaries. The drug was banned. Further research has led to the use of thalidomide to treat leprosy and cancer, though it is still not suitable for pregnant women. Leprosy can cause skin and nerve damage, muscle weakness and permanent disability. Sufferers of leprosy can be rejected by the community and cannot work. As a result there is a 'black market' for the drug thalidomide and it is taken without medical guidance. Some women in Mexico have used it while pregnant and have given birth to children with abnormalities.

Test yourself

13 Describe the side effects of the drug thalidomide. Why do you think the side effects of thalidomide were not detected when the drug was tested?

14 Would it be right to test a new drug on a pregnant woman? What is a possible alternative?

15 Give three positive points about statins.

16 How do statins work to reduce cardiovascular (heart) disease?

Testing drugs

HOW SCIENCE WORKS
ICT

You have developed a drug that can make people more intelligent. Create a multimedia advertising campaign explaining what your drug can do and showing that it has been properly tested and is safe to use.

What are statins?

In Chapter 1 we saw that cholesterol is produced naturally in the liver and that some people produce too much. You will remember that too much cholesterol increases the build up of fatty deposits in artery walls and increases the risk of heart disease. A large amount of research has provided evidence that lowering levels of 'bad cholesterol' in the blood significantly reduces the risk of heart disease and heart attacks (studies show by about 25%). The drug statin reduces the production of 'bad cholesterol' by the liver and so reduces blood cholesterol levels. Statins are not costly and are among the safest drugs. They are used for people who are at risk because they have high cholesterol and a history of heart disease in their family.

B1 2.8 Recreational drugs

Learning outcomes

- Understand the possible progression from non-addictive recreational drugs to hard drugs.
- Evaluate the claims of researchers about the potential risks of cannabis.
- Evaluate the ethical and medical implications of the use of drugs in sport.

If a drug is legal does that mean it is safe?

The simple answer is no. Caffeine, alcohol and cigarettes are legal but every year more people die as a result of drinking alcohol or smoking than die from illegal **recreational drug** use. Prescription medication, such as strong painkillers or tranquillisers, is often **misused** by people who take it for its mood-altering effects, with fatal results. 'Legal highs' are drugs or herbal preparations (herbal ecstasy) made to be chemically similar and give the same effects as banned drugs. Unfortunately, they have not undergone the long safety testing of medicinal drugs and sadly some users die of the effects. Do you think they should be made illegal?

Recreational drugs are used because of the effect they have on changing a person's mood, emotion or state of consciousness.

- Stimulants, such as cocaine, make people feel full of energy (but cocaine can cause heart failure).

A **recreational drug** is a chemical substance that acts on the brain and nervous system, changing a person's mood, emotion or state of consciousness.

Drug misuse is when a person takes one or more drugs to change their mood, emotion or state of consciousness.

- Depressants (or sedatives), such as heroin, make people feel relaxed (but heroin can cause respiratory failure or loss of normal lung function).
- Hallucinogens, such as ecstasy, make sounds and colours more intense and give the feeling of being able to dance all night but also reduce the body's ability to regulate temperature. Dancing in a hot club without this internal temperature control results in the body seriously overheating (up to 43°C), so hyperthermia sets in. Ion balance is also disturbed and might result in blood-clotting.

Everyone thinks 'it will never happen to me' and 'everyone else is using' — and then someone collapses and is rushed to hospital. Two points to remember are (i) drugs bought on the street are often diluted with other chemical substances (even rat poison and cleaning powders have been used!) and (ii) we are all different — think about allergies — so we react differently to medication and drugs. The following website gives some real-life experiences: http://www.nhs.uk/conditions/Drug-misuse/Pages/Introduction.aspx

Test yourself

17 a Name three different types of illegal drugs.
b Why do some people use recreational drugs?
c Medicinal drugs go through careful testing and packaging. How do 'street drugs' differ?

Cannabis

Users claim that cannabis makes them feel relaxed and happy and it increases sensations induced by colour and music. Short-term effects include day-dreaming and the inability to concentrate or organise facts. Tests on people using a computer simulator showed that more errors were made 2 hours after smoking. Because this is a readily available street drug there has been much research into the smoking habits of cannabis users to investigate longer-term effects, such as whether or not there is a link:

- between regular use and moving on to hard drugs
- between smoking cannabis and development of mental health problems, such as schizophrenia

Recent research has shown that, for the majority of users, cannabis is not a 'gateway' drug that leads to hard drugs. However, a minority who already have behavioural problems and who start using cannabis early do move on to hard drugs. The majority of heroin users started with alcohol, cigarettes and cannabis.

Evidence from 30 studies shows that:

- between 18 and 30% of schizophrenia cases are linked with cannabis
- heavy users of cannabis at age 18 were 600% more likely to be diagnosed with schizophrenia in the next 15 years compared with non-users
- people using at age 15 were 300% more likely to develop schizophrenia than non-users, and 700% more likely to develop other mental illness

These findings are based on thousands of people and involved follow-up studies over many years.

Test yourself

18 Give two reasons why it might not be a good idea to accept a lift in a car from someone who was smoking cannabis earlier in the evening.

19 Give two possible long-term problems resulting from regular cannabis use.

Additional biological facts:

- During teenage years the neural connections in the brain are 'reorganised' and so they are more vulnerable to disturbance up to about age 20.
- The strength of the active component in cannabis has increased, particularly in skunk and synthetic cannabis.
- Cannabis smoke is more harmful than cigarette smoke as it contains many chemical compounds.
- Cannabis has features of an addictive drug.

Addiction

Although it is true that for most people the initial decision to take drugs is voluntary, one of the biggest risks of drug misuse is developing drug addiction. Frequent abuse, even regular smoking of cannabis, leads to changes in the nerve connections and chemical reactions in the brain. Regular abuse changes the functioning of the brain, which can affect a person's self-control and decision making and cause drug craving. Heroin and cocaine are very quickly addictive. There are two main types of drug addiction:

- physical addiction — if the supply of the drug is suddenly withdrawn there are withdrawal symptoms, such as vomiting or cramping
- psychological addiction — there is a craving for the drug to get the 'high' or 'relaxation' effect. If the drug is withdrawn, there may be no physical symptoms but there may be psychological symptoms such as depression, anxiety and irritability.

Test yourself

20 How do some drugs cause addiction?

Performance-enhancing drugs in sport

Most banned substances can be detected in urine samples. Those that are close copies of natural hormones can also be detected from side effects.

Why do some athletes use anabolic steroids? Steroids stimulate muscle development when used in regular training with a high protein diet. Their use reduces recovery time after training, meaning that more time can be spent training. In a race, steroids increase aggressive competitiveness, and increase strength and speed over short distances.

Athletes testing positive for any banned substance are disqualified. If they are not 'found out' is it a 'fair test'? Apart from cheating, is there a good reason for banning drugs or should everyone be allowed to use them?

Allowing steroids would expose athletes to the side effects, which are liver damage, aggressive mood swings that might include violence, infertility and baldness in men, and hair growth and a deep voice in women. There are other illegal performance-enhancing drugs but all have side effects which harm the athlete. If allowed, not all athletes could afford the drugs and there are risks with impure products. There will be a great deal of time and money spent on drug testing at the 2012 Olympics.

Figure 2.20 Marion Jones won five medals at the 2000 Olympics but confessed to drug taking and was stripped of her medals. Look at her muscles.

HOW SCIENCE WORKS ACTIVITY

You work for WADA (the World Anti-Doping Agency) and need to develop a drugs testing programme for a sport of your choice.

What drugs would you be testing for?

Why would these drugs give a performance advantage to the participants?

How would you test for them?

Effects of drugs

HOW SCIENCE WORKS ICT

Using the website www.talktofrank.com, create a conversation between two people — one a drug user, the other working for a drugs advisory service on one chosen drug.

The conversation needs to include the effects of the drug, why the user takes it and what the negative effects are, and how the advisor might try to get the user to stop.

You can do this as a script, a play, a radio podcast, a video or an animation using a site such as www.dvolver.com.

Homework questions

1 A student accidentally touches a hot saucepan. His hand automatically moves away from the pan.

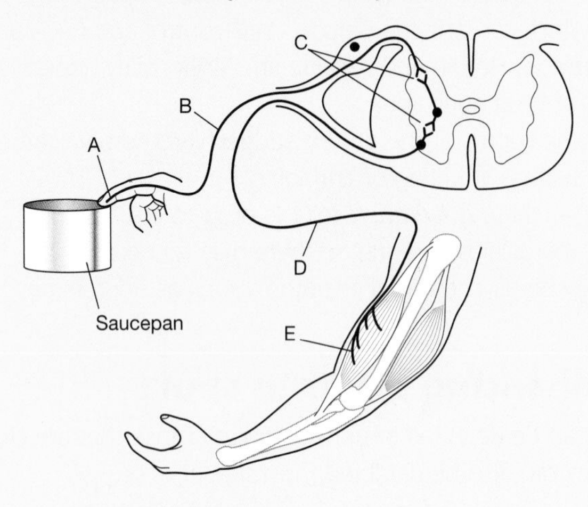

a In this reflex action:
  (i) where is the receptor?
  (ii) where is the effector?

b Explain how an impulse crosses the synapse labelled C.

c Explain why this type of reflex action is important to our bodies.

2 A hormone is involved in controlling the water content of the body.

a State two ways in which water leaves the body.

b What can cause a rapid fall in the water content of the body?

3 Read the information about contraceptive implants:

A contraceptive implant works a bit like a contraceptive pill. It contains a hormone that is also present in contraceptive pills which prevents pregnancy. The hormone implant is a small, thin flexible rod, 4 cm long and made of plastic. It is inserted just under the skin of a woman's arm. This procedure must always be undertaken by a doctor who is familiar with the technique. The implant gradually releases a small amount of the hormone, which prevents pregnancy for up to 3 years.

a How can a hormone prevent pregnancy?

b Give one drawback of using contraceptive implants rather than contraceptive pills.

c Contraceptive implants are being used increasingly in birth control programmes in developing countries, instead of contraceptive pills. Can you think of a reason for this?

4 Medicines must be thoroughly tested before they can be prescribed to the public.

a Outline the procedure for testing a new medicine.

b A drug called thalidomide caused severe side effects despite passing safety tests and clinical trials.
  (i) Describe the side effects caused by thalidomide.
  (ii) Explain why thorough testing did not identify the possibility of these side effects.

5 A newspaper suggests that the costs of major sporting events would be reduced if there was no drug testing.

a Are there any scientific reasons for banning drugs?

b Is it any more unfair to use steroids than a low-friction swimming costume?

c Suggest some ethical reasons why performance-enhancing drugs should not be used.

6 Statins reduce the amount of cholesterol in the blood.

a Why is this reduction advantageous?

b What types of illness would statin use reduce?

c Which people will benefit most from this treatment?

d Statins are cheap to manufacture and have few side effects. Suggest reasons for and against giving them to everyone over 40 years old?

e If a person has high cholesterol are there any lifestyle changes that they could make to reduce the problem?

Exam corner

A plant was grown in a pot. The pot was then knocked over. The shoot turned and grew upwards as shown in the photograph. Explain in detail what caused the stem to bend upwards. *(6 marks)*

Examiner comment This type of question tests your ability to write in extended prose. The answer should present facts correctly in a logical sequence. It is a good idea to jot down the keywords under the question as you plan your answer.

Student A

The plant leaves wanted light ✗ so the plant bent up ✗ so the light would shine on the leaves. ✓

Examiner comment Student A earned 1 mark for giving the result of the phototropic response.

Student B

A plant hormone ✓ was produced at the tip ✓ of the stem. It went backwards to the target region ✓ and here the cells at the bottom of the stem got longer ✓.
The stem bent up showing a positive geotropic response. ✗

Examiner comment Student B made a good start. He identifies the compound and gets the mark for 'hormone', even though he does not use the term auxin. He has given the site of production and mentioned target or sensitive region. He does not state that the auxin accumulates at the lower side but does state that the cells get longer. Unfortunately he misnames the response — it should be negative geotropism. He does not give the importance of the response.

Chapter 3 Interdependence and adaptation

Setting the scene

What causes a mouse population to get out of control? What factors would normally control a population? How can mice reach plague proportions so quickly? In this chapter we look at population factors and even consider 'life on Mars'.

ICT

In this chapter you can learn to use a range of resources to show what you know about how organisms are adapted to local conditions.

Practical work

In this chapter you can learn to:

- make predictions about the conditions for decay
- look for patterns in data and draw conclusions

B1 3.1

Learning outcomes

- Know how to identify special adaptive features of animals.
- Understand how adaptations allow an animal to survive in hostile environments.
- Explain the adaptations of plants for different environments.
- Explain the adaptations of microorganisms for different environments.

Figure 3.1 Polar bears are adapted to survive in cold conditions.

Figure 3.2 The Arabian camel is adapted to survive in hot, dry conditions.

Adaptations

At Key Stage 3 you investigated food chains, food webs and 'who ate whom'. In this chapter we will look in greater depth at the adaptations of organisms to the specific environments in which they live and at the factors which control population size.

Animals are said to be adapted to their surroundings when they are able to survive and reproduce successfully. In the case of the mouse population, the environment provides a supply of food for growth and a nest site or place to raise their young.

If you were going on a field trip to the Arctic, what clothing would you take? You would probably take a thick fleece and windproof jacket. Animals that live in the Arctic must also be able to withstand the cold conditions.

Figure 3.1 shows the features or **adaptations** that enable a polar bear to survive in the Arctic. The polar bear has:

- a small head and ears, giving the minimum surface area for heat loss
- a compact body shape, which is also streamlined for swimming
- a thick layer of fur — this traps air, which is a good insulator
- a thick layer of fat (up to 11 cm thick), which insulates against heat loss and also acts as a food reserve during the Arctic winter when the bear is in a deep sleep
- white fur, which camouflages the bear, enabling it to get closer to prey when hunting
- white fur, which reduces heat radiated from the body

Boost your grade

Read through this list again. It is not *just* a list — for every feature an advantage is stated. For example, thick fur traps air, which is an insulator. Learn these points together.

An adaptation is a feature that allows an organism to survive in the environment in which it normally lives.

An arid region is hot and dry. This results in sparse vegetation which is usually adapted to reduce water loss and attack by herbivores.

A camel lives in hot **arid** conditions. It needs to prevent overheating and is adapted to this environment by having:

- long legs and neck, giving a large surface area for heat loss
- thin hair on top of the body to allow heat loss
- no hair on the underside of its body, making heat loss easier
- little body fat so heat is easily lost from the skin capillaries
- a fatty hump that can act as a food reserve when desert food is scarce
- camouflage colouring
- two rows of eyelashes for protection from sun and sand
- nostrils which close for protection during sandstorms

Compare the camel's features with those of the polar bear.

Test yourself

1 Look carefully at the photographs of two different foxes and answer the questions below.

Figure 3.3 The fennec fox is found in the Sahara Desert.

Figure 3.4 Arctic fox.

a Which animal is adapted to retain body heat?
b Explain how its body features help to retain heat.
c How do the colours of their fur relate to their environment?

2 The Arctic fox moults its coat in the spring. What colour would you expect the summer coat to be?

3 What is the purpose of the thick layer of fat on the Arctic fox?

4 The Arctic fox has short legs covered with thick fur. What would you expect the legs of the Fennec fox to look like?

5 Why do the two foxes, which both listen for prey, have such different sized ears?

Figure 3.5 A house leek showing the fleshy leaves with a waxy coating.

Many plants are also adapted to live in harsh environments with features to stop them losing too much water. The house leek (Figure 3.5) lives on rocky outcrops where rainfall amounts vary during the year, with none at all during the summer. Similar conditions are found when they are grown on roofs or in rock gardens.

Water loss is reduced by having:
- a short stem
- fleshy green leaves that store water but dry up at the end of the year
- a waxy, shiny outer covering to the leaves
- long roots that penetrate deep into the soil in the rock crevices

Some organisms also have adaptations to prevent animals eating them.

Figure 3.6 Look at the adaptations of the thorn bush. What adaptations do you need to eat thorn bush leaves? A long prehensile tongue up to 45 cm to strip the leaves while avoiding the thorns, and a leathery mouth.

Figure 3.7 The cane toad has poison glands that produce a toxin which stops the heart. What do you think happens to any predator?

Figure 3.8 What visual stimulus makes you avoid wasps?

Figure 3.9 This is a harmless hoverfly. Why might insect predators avoid this?

From these examples you can see that plants and animals can survive in different conditions because of special adaptations that suit their environment. These may be specialised structures or different enzymes.

Extremophiles

A **methanogenic bacterium** is described as an **autotrophic** hyperthermophile, which means that it can make its own food energy and survive very high temperatures. It has heat-stable enzymes and derives its cellular energy from hydrogen and carbon dioxide. It releases methane. There is no oxygen in this environment. Speculation suggests that these bacteria might have been some of the first organisms when Earth was a young planet and life forms like these might be found on other planets.

Figure 3.10 Could anything survive in a hydrothermal vent with temperatures close to 100°C and pressures high enough to crush a submarine?

A **methanogenic bacterium** lives in an anaerobic environment and produces methane gas. (You will also come across methanogenic bacteria in anaerobic digestion in waste disposal and digestion in the rumen of a cow.)

An **autotroph** is an organism that can make its own food from inorganic sources. For example, plants make glucose from carbon dioxide and water using energy from sunlight; some bacteria make glucose from carbon dioxide and hydrogen sulphide. They obtain the energy by breaking down the H_2S.

Extremophiles are organisms with adaptations that enable them to live in conditions that are far outside normal ranges — for example, extremely high temperatures, high salt or freezing cold.

Halophiles are microbes that can live in very salty environments such as the Dead Sea and the Great Salt Lake. They use oxygen for respiration. An obvious feature is the pink coloration from the pigment carotene, which prevents ultraviolet damage. Halophiles also have a specialised photosynthetic pigment and specialised cell walls with a protein-like coating. They are able to store potassium compounds, which allows them to balance the salts inside the cell with those outside to prevent water loss. Might these be found on Mars?

Test yourself

6 What special adaptations would you expect in bacteria found in volcanic hot springs?

7 How do methanogenic bacteria obtain energy when there is no oxygen for respiration?

Figure 3.11 Salt flats. Why are they red?

NASA scientists had a surprise in 2005 when they melted some ice from a frozen Alaskan pond and found a new species of bacterium swimming on the microscope slide. It had been frozen dormant for 32 000 years — from the time when woolly mammoths were around.

Test yourself

8 List the adaptations found in microbes that can live in very salty conditions.

Is there any value in studying bacteria from extreme conditions? Bacteria have been used for many years to break down waste in municipal tips and wastewater treatment. By carefully selecting bacteria from some extreme environments, cultures of bacteria can be produced and modified by genetic engineering that will break down specific hazardous pollutants from industrial processes. Industrial uses include breakdown of toxic organic compounds such as phenol and PCBs, breakdown of oil after tanker spills or the recovery of uranium from low-level radioactive waste.

Local adaptations

HOW SCIENCE WORKS
ICT

Produce a resource showing the adaptations of a plant or animal that lives in your school grounds. You need to answer these questions:

- What are the threats to the survival of the organism?
- What is it competing for, and with what is it competing?
- How is it adapted?
- Produce a labelled picture, a report, presentation, video or podcast to show your findings.

B1 3.2

Learning outcome

Understand competition among plants and animals.

What determines the number of organisms in an environment?

Plants compete for light to photosynthesise and for soil space from which to absorb water and take in minerals. This interaction of organisms (plants or animals) trying to obtain the same food or occupy the same territory is called **competition**.

Consider a hedgerow beside a road. Plants such as primroses, like those in Figure 3.12, flower early in the year to avoid competition for light. They produce leaves, flowers and seeds before the tree leaves open and put them in shade.

Animal populations are also regulated by competition among members of the same species and competition between different species. Some animals and birds compete for a territory of sufficient size to feed their offspring. Although people love to hear blackbirds and robins singing, their songs are actually 'war cries' to keep other males off their territory. These birds sing from different high points around the area to mark the boundary. Can you name any other animals that have a territory? How do they mark it?

Figure 3.12 The primroses in this hedgerow flower before the trees open their leaf buds.

Test yourself

9 What do plants compete for?

Competition is the interaction between organisms trying to obtain the same food or occupy the same territory.

A predator is an animal that catches and eats another animal. The animal caught is called the prey.

Gannets are sea birds that catch fish by diving head-first into the water. They live and breed on remote rocks or cliffs. Just imagine the noise in the colony when all the birds are competing for mates and nesting sites. Notice in Figure 3.13 that the nests are placed 'pecking distance' apart. You can see why when you look at the sharp, pointed beak of the gannet!

Figure 3.13 A gannet colony. Why are the nests so evenly spaced?

The size of a gannet population changes with food availability. If more fish are available, more young gannets are raised, but this increases the competition for nest sites in future years. Other limiting factors within the gannet colony are **predators**, such as gulls, who brave the gannets' spear-like beaks to steal eggs or young.

In general, the number of organisms in an environment is strongly influenced by food supply. Numbers will increase if there is plenty of food and decrease if food is in short supply. From the examples above, you can see that population size is also determined by competition among members of the same species, and by competition with other species for the same nutrients or space. This applies to both plants and animals.

Figure 3.14 A gannet will use its large powerful beak to attack any object or bird that strays into its territory

Test yourself

10 For what resources do animals of the same species compete?

11 Owls and kestrels compete by hunting for small rodents for food. How do they reduce competition?

B1 3.3

Learning outcomes

- Know that lichens and invertebrate animals can be used as indicators of pollution.
- Understand that environmental change can be measured using non-living indicators such as oxygen levels, temperature and rainfall.

A **biological indicator (bioindicator)** is a plant or animal that can indicate the level of a particular chemical in its habitat. In the examples in this section sensitive lichens indicate sulfur dioxide levels in the air and invertebrates indicate oxygen levels in water.

Living organisms as indicators of pollution

Chemical analysis of water or air for pollutants gives a measure of what is present at the time of sampling. However, it cannot give information about pollution incidents that may already have been diluted or blown away. Such incidents may kill certain species, leaving evidence for the biological detective.

Pollution indicator species

Different species of lichens have different sensitivities to sulfur dioxide (Table 3.1). Some are so sensitive that even a trace of the gas in the air will kill them. This means that we can use the lichens that are still growing as bioindicators to find out the prevailing levels of sulfur dioxide pollution in the air.

Information from a lichen survey could help a local authority to monitor changes resulting from increased traffic density or industrial activity. This would enable it to make decisions about locations for new roads or industrial sites.

Lichens are pronounced 'li kens'. They are a mutual association of a fungus and an alga. They occur as crusty patches on rocks, tree trunks and walls. Lichens are living organisms that can be used as **biological indicators** (**bioindicators**) of pollution as the algal part is rapidly killed by sulfur dioxide in polluted air. Lichens grow slowly over many years.

Table 3.1 Common lichens which act as bioindicators of air pollution from sulfur dioxide.

Type of lichen	(1) Common orange lichen, quite tolerant of moderate pollution levels	(2) Quite a common lichen but dies quickly if pollution levels rise	(3) A common woodland lichen and an indicator of clean air	(4) Beard lichen, only survives in pure air
Maximum level of sulfur dioxide tolerated	70 μg/m^3	60 μg/m^3	50 μg/m^3	35 μg/m^3

Carrying out a pollution survey

HOW SCIENCE WORKS
PRACTICAL SKILLS

A science class decided to survey the air quality around their town, using the above lichens as indicators.

Starting hypothesis: The lichens growing will be those able to survive in the level of sulfur dioxide pollution found normally at that location.

Prediction: As sulfur dioxide is produced by combustion of fossil fuels, the levels of the gas will be highest near main roads, and residential and industrial areas. Therefore, only tolerant lichen will be found in these areas. The lowest concentrations of sulfur dioxide will be found on the southwest side of the town in the old woodland and farmland areas and so more sensitive lichen will be found here.

Procedure: Before starting all the students made a copy of the town map on which they marked industrial areas in blue, coloured the main roads in red and all parks, woodland and fields in green. From the town centre they drew eight lines along the main compass bearings (north, northeast, east, southeast and so on) and made concentric circles from the town centre 0.5 km apart, up to a 5 km radius. Survey points were chosen where the compass lines and circles crossed. Anyone living near an intersection surveyed that location. More distant or difficult points were visited by a group with their teacher. They were told to look on the nearest walls, trees, or hedges to find lichens that were a close match to the photographs of the indicator species.

On return to the laboratory, the results were plotted on the map.

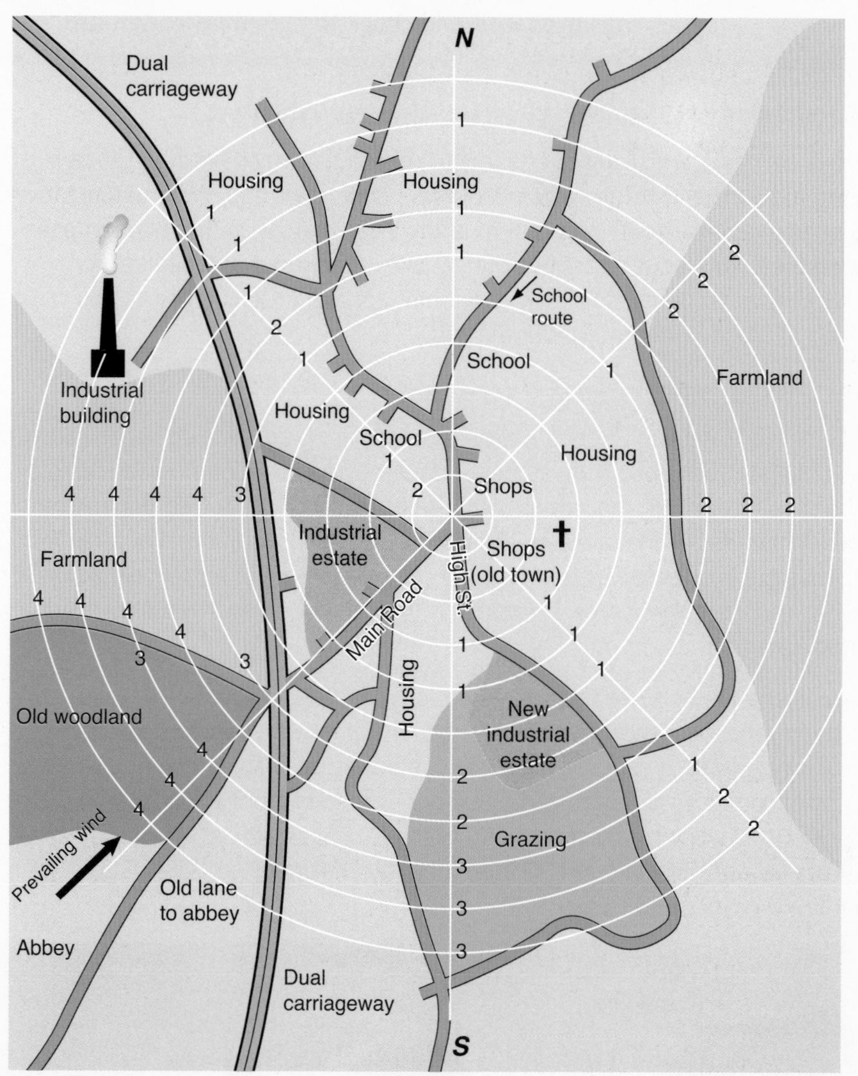

Figure 3.15 Map showing the results of the lichen survey.

"chemical spraying on farmland might make a difference"

"the sizes and ages of the trees surveyed were different"

"some areas were newly developed and had no lichens at all"

"more types of lichens were found than given on the pictures"

"it is not easy to separate the variables in this invesigation"

"not enough results to draw a definite conclusion"

Figure 3.16 The class evaluated their results together. Here are some of their comments.

1. Which gas are lichens most sensitive to and what produces this gas?
2. a What was the direction of the prevailing wind?
   b How could this affect the distribution of sulfur dioxide?
3. a Where were the most sensitive lichens found?
   b Did this fit the prediction made?
4. Students reported finding dead and dying lichens in the hedge beside the road running southeast from the old town. What does this suggest?
5. a Where were the least sensitive orange-coloured lichens found?
   b Look at the map and give reasons to support your answer.

6. The churchyard in the town centre had a variety of grey and orange lichens but none was found around a newer church to the east of the town. Suggest why.
7. What instructions were given to students going to a survey point?
8. Using the map, write a conclusion, stating what the survey found.
9. Do the results of this investigation support the prediction made?
10. Using the points made by the group and any others of your own, comment on the **validity** and **reliability** of this investigation.

If you undertake an investigation of this kind, you must follow the risk assessment that your teacher has prepared and check the weather conditions.

Reliable results are results you can trust and that can be reproduced by others.

Valid results measure what they are supposed to be measuring. They must be obtained by a fair and unbiased test.

Figure 3.17 Brown trout are sensitive to pollution.

Figure 3.18 What is this sampling method called?

Bioindicators for water oxygenation

Small animals, particularly invertebrates and pollution-sensitive fish, can be used as bioindicators for water cleanliness. Some years ago, there was great excitement when trout were caught in the River Thames as they had not been seen there for many years. But trout are not usually used as bioindicators in checking water purity; smaller invertebrates are more useful. They can be caught easily in shallow streams (see Figure 3.20).

Normal sampling method

The normal sampling method is kick-sampling, using a D-shaped net to catch the invertebrates disturbed from stones and plants. The investigator wades across the stream, shuffling the stones with his/her feet. The net is held downstream beside the feet, in the direction of flow, to catch the specimens. For a fair test, the sample is collected by shuffling the same distance and for the same time when each sample is taken. The catch is then emptied into a white tray with a little water in the bottom.

Specimens can be sorted using a teaspoon, pipette and paintbrush. Each different species is recorded as rare, frequent or in large numbers. A clean well-oxygenated stream will have a high biodiversity (many different species). A polluted stream will have a low number of species that have adaptations to survive in low oxygen conditions. An example of such species is chironomid midge larvae, which are red because they have haemoglobin to pick up oxygen in poorly oxygenated water.

Figure 3.19 These students are sorting through a kick-sampling catch.

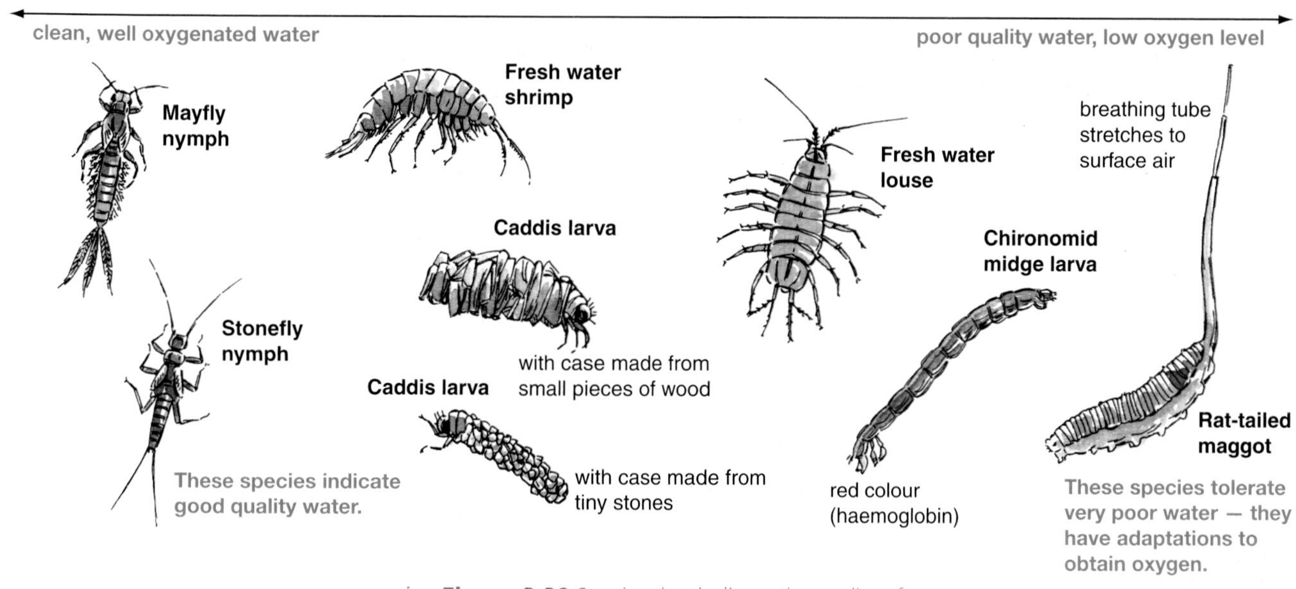

Figure 3.20 Species that indicate the quality of water.

Figure 3.21 A simple rain gauge. Look at the units. The scale is graduated to give the reader a direct measurement in useful units for comparison with other sites or on different days.

Measuring non-living indicators of environmental conditions can also be carried out using simple apparatus.

Dissolved oxygen can be measured using an environmental oxygen electrode which is attached to a meter for the read-out. You dip the probe into the water and wait until the value on the meter is constant. You could use this to measure how the dissolved oxygen level varies during the day in a pond as the pondweed photosynthesises, or you could compare the biodiversity of a well oxygenated pond with one with a low dissolved oxygen content.

Temperature can be measured with a simple thermometer or a maximum and minimum thermometer, or by using a temperature probe attached to a meter. You could compare temperatures in a compost heap that was showing good decomposition with one where there appeared to be little change.

Figure 3.22 A dissolved oxygen electrode and meter.

Local pollution

HOW SCIENCE WORKS
PRACTICAL SKILLS

Where might pollution in your local stream or river come from? How could you find out the source of the pollution and the effects of the pollution on the animal and plant life in the stream?

Chapter 4
Genetic variation and evolution

Setting the scene

Animals evolve; ideas evolve. In this section we see how some of the techniques that were science fiction last century are now part of everyday technology. What does the term genetic modification mean to you?

ICT

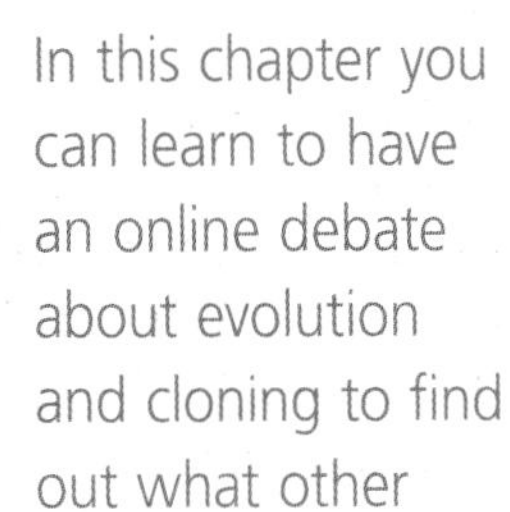

In this chapter you can learn to have an online debate about evolution and cloning to find out what other people think.

Practical work

In this chapter you can learn to work safely when cloning plants.

These calves were produced by embryo transfer, the Humulin and golden rice by GM technology.

B1 4.1 Interpreting the evidence for evolution

Learning outcomes

- Evaluate different theories of evolution.
- Explain why Darwin's theory of evolution is now the most widely accepted.

This chapter almost begins at the beginning. It follows the development of ideas and how they have lead to modern genetic applications.

We live on a planet with a fantastic variety of different animals and plants. Scientists have estimated that there are between 2 million and 100 million different species of plants and animals on the Earth today, but they don't know the exact number. How did all these different animals and plants come to be living on Earth? Have there always been similar plants and animals on Earth? Are the present plants and animals like their predecessors (ones that were alive in the past) or very different from them? Although there are answers to some of these interesting questions, scientists cannot be certain how life began on Earth.

Most scientists believe that the plants and animals alive today have evolved from forerunners that lived in the past. These forerunners were not the same as today's species. Changes have taken place over many generations. Scientists call the overall process evolution. Evolution has resulted in species that are adapted to survive in the environments in which they live, as we saw in Chapter 3. No one has actually seen these changes taking place in animals because the process takes so long, but changes have been followed in bacteria. There have been many theories that try to explain evolution, but Darwin's theory is the most widely accepted.

Figure 4.1 Some of the species of animals and plants found on the Earth today.

Conflicting theories of evolution

The **fossil** record provides evidence of change over very long periods of time. Lamarck and Darwin suggested different **hypotheses** to explain these changes.

A **fossil** is the remains or imprint of dead plants or animals trapped in sedimentary rocks when the rocks were forming millions of years ago. Animal remains have been mineralised and turned into stone.

A testable statement is called a **hypothesis** (plural: hypotheses). In some cases, if it can be tested and proved correct it becomes a theory, and this will be further tested.

Lamarck (1744–1829) was a French natural philosopher. Lamarck's theory was related to man and was meant to please the politicians. He suggested that a desire or need for change caused that change to happen in the organism during its lifetime and this would be passed on to the offspring. He stated that structures that are constantly in use are well developed and those characteristics acquired during life would be passed on, i.e. 'transmission of acquired characteristics results in evolutionary change'. For example, the labourer needed big muscles to do his job and he would pass these on to his son. Lamarck's 'evidence' was that when the environment causes the need for a structure, this

3 Figure 3D shows the carbon cycle.

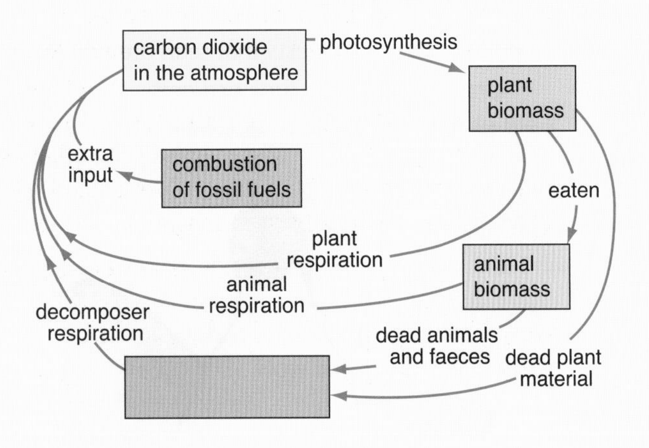

Figure 3D

a Name the only process that removes carbon dioxide from the atmosphere? ***(1 mark)***

b What structural compound is made by plants from the carbon dioxide? ***(1 mark)***

c What organisms belong in the brown box? What is their role? ***(2 marks)***

d Fungi break down dead organic matter by secreting enzymes. A mouse has died; name one enzyme that will be needed to break down the body. ***(1 mark)***

e Give two ways in which the breakdown products could be used. ***(2 marks)***

Chapter 4

1 Young rabbits, like these, often look like their parents. This is because information about their appearance, for example fur tone, is passed from parents to their offspring.

Figure 4A

a How is information passed from the parents to the offspring? ***(1 mark)***

b What is the name for this type of reproduction? ***(1 mark)***

c In the nucleus of the cell are chromosomes. In what form is the genetic information on the chromosomes? ***(1 mark)***

d Rabbits can suffer from a virus disease called myxomatosis. Some rabbits are immune to this disease. Suggest how this might have come about by natural selection. ***(3 marks)***

2 Humans reproduce by sexual reproduction. Michael and James are brothers but they have very different features.

Figure 4B

a State two ways in which sexual reproduction is different from asexual reproduction. ***(2 marks)***

b Explain why Michael and James have different features. ***(3 marks)***

3 Read carefully the following text:

A woman from Texas ordered a genetic replica of her cat, Nicky, when he died aged 17. The new cloned cat, named Little Nicky, cost his new owner

$50 000. He was created using DNA taken from his namesake Nicky. 'He is identical. His personality is the same,' said his owner. During an interview, the new owner asked that her surname and hometown were not made public. She feared that she may become a target for groups that oppose the use of cloning techniques.

a Explain why Little Nicky is exactly the same as Nicky. *(3 marks)*

b Little Nicky's owner is concerned that she may become a target for anti-cloning groups.

(i) Give one example of a group or organisation that may disagree with the cloning of pet animals. *(1 mark)*

(ii) Why do you think such groups disagree with these cloning techniques? *(1 mark)*

c What biological problems might this product of cloning experience? *(1 mark)*

4 a Provide an explanation for the theory of evolution. *(2 marks)*

Figure 4C

b The following text provides a possible explanation by Lamarck for the evolution of the long legs of the heron:

When an animal undergoes change during its lifetime and then mates, it passes this change on to its offspring. For example, herons did not always have long legs. Herons gradually developed long legs after their ancestors started to feed on fish. As they walked into deeper water, they would stretch their legs to prevent their bodies from becoming wet, causing their legs to lengthen. Their new trait of longer legs would be passed on to their offspring, who would also stretch their legs. Over time, the legs of these wading birds became longer and longer.

Darwin would have provided a different explanation for the evolution of the long legs of these birds. What are the main differences between his explanation and that of Lamarck? *(3 marks)*

Biology 2

Chapter 5
Cells, tissues and organs

Setting the scene

Cells are the basis of life. The DNA in cell nuclei controls all the processes in your body that keep you alive and is responsible for reproduction. The cells in the pictures are all different and each is specialised for a specific function. Can you describe the function of the cells in the photos? Roughly how big is each cell? Can you remember the names of any of the structures inside cells?

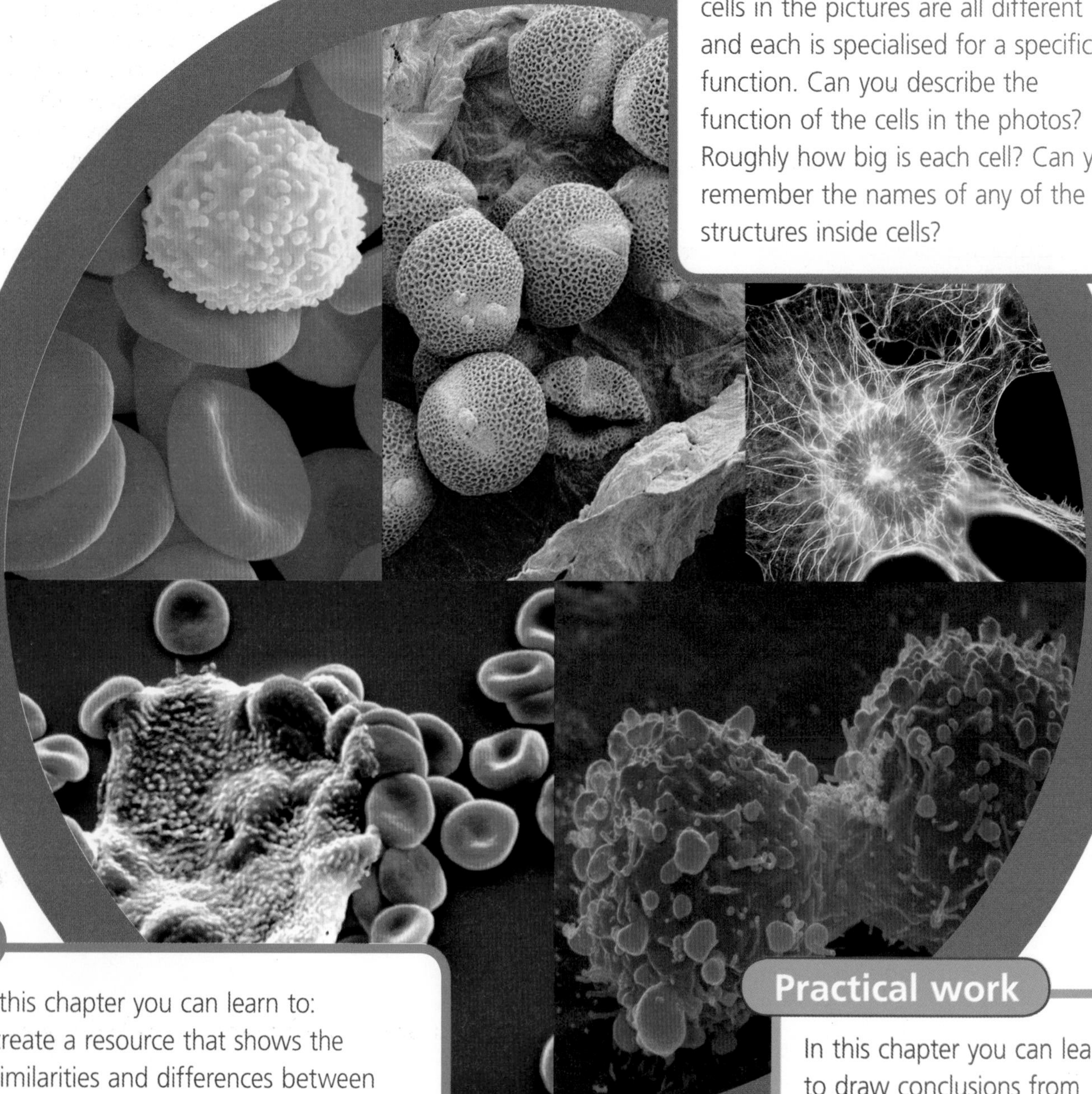

ICT

In this chapter you can learn to:

- create a resource that shows the similarities and differences between plant, animal, yeast and bacterial cells
- use multimedia to explain diffusion

Practical work

In this chapter you can learn to draw conclusions from the results of a diffusion experiment.

B2 5.1

Learning outcomes

- Know the names and be able to explain the function of the different parts of animal and plant cells.

How do cells perform different functions?

In this section we look at the component parts of cells and their role in cell function.

Today we understand that all living things are made up of cells, but this was not always the case. In 1665 a scientist called Robert Hooke used a primitive microscope to discover that samples from a cork tree were made up from repeating structures. He called these structures cells.

Hooke's discovery was only the start of what is now known as cell theory. Although he observed cells in one type of plant, he did not link cells to other species and had no idea what was inside cells.

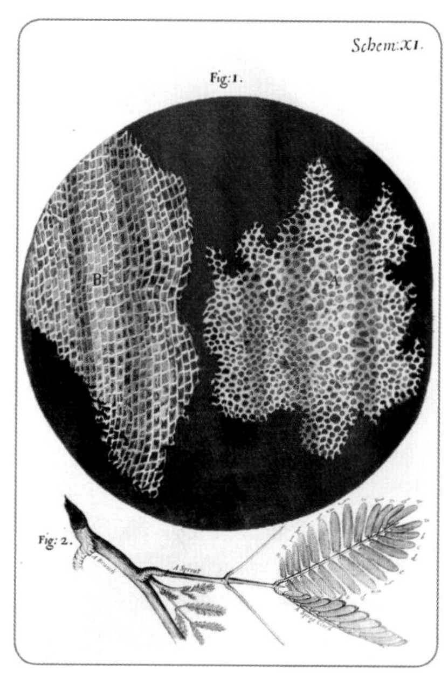

Figure 5.1 Robert Hooke's drawing of the cork cells he could see with his microscope.

In 1831 improvements in microscope technology allowed a scientist called Robert Brown to observe nuclei inside a number of different cells. Later in the same decade two German biologists, Theodor Schwann and Matthias Jakob Schleiden, carried out more research and linked cells to all plants and animals. Schwann published these findings without any acknowledgment of Schleiden's or anyone else's contributions. He was correct to suggest that cells are the units that make up all living things. However, he also suggested that new cells formed out of nothing. This was proved to be wrong by Rudolph Virchow in 1858. Virchow proposed the theory, still in place today, that all cells are formed from other cells.

Test yourself

1. Cells are often described as the 'building blocks of life'. What does this mean?
2. Robert Hooke observed cells but did not identify anything inside them. What was limiting his research?
3. Why do you think that Robert Brown's discovery was important in developing a better understanding of cells?

What are the functions of different parts of a cell?

Modern microscopes magnify thousands of times and now let us see an incredible amount of detail about the structure of cells — detail that Robert Hooke's microscope simply could not display. We now know that even within cells there are structures that have specific functions. We can identify different parts of cells that each carry out different functions. From our observations we have discovered that all cells have some parts in common, such as a **cell membrane**. We will look at each structure and its function.

Animal cells

Nucleus

The **nucleus** of a cell contains an individual's genetic information as genes on the chromosomes.

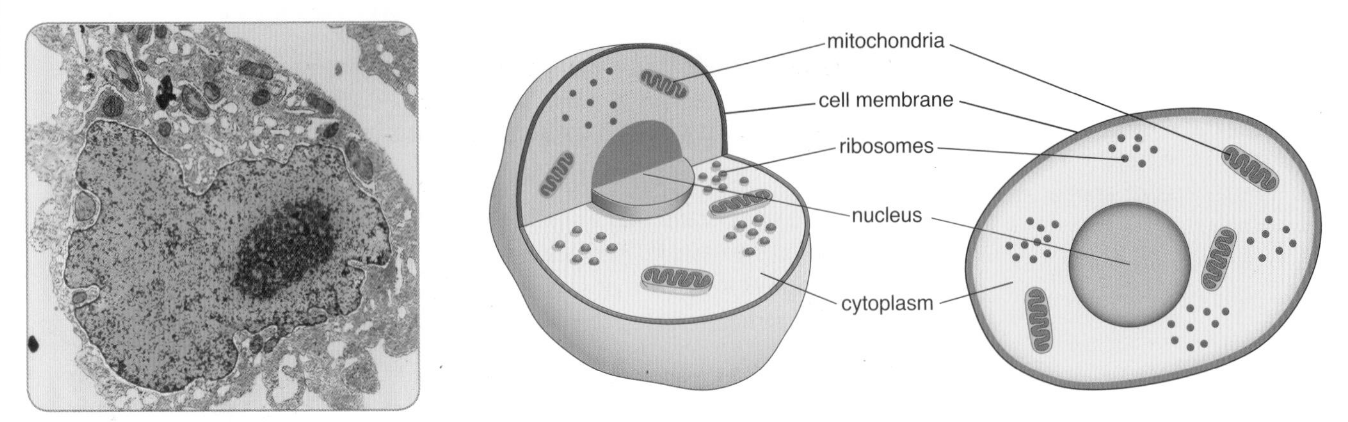

Figure 5.2 An animal cell shown as it is seen with a microscope (magnified × 4800) and as three- and two-dimensional diagrams.

The **nucleus** controls the cell chemical reactions and cell division.

The **cell membrane** surrounds the cell and controls the movement of substances into and out of the cell.

The **cytoplasm** is where the chemical reactions of the cell take place.

The nucleus has two main functions:

- it regulates the cell's activities by controlling the production of enzymes, and
- it controls cell division

How does the nucleus regulate the cell's activities? The chromosomes in the nucleus carry the genetic code for production of all proteins including enzymes. The code is copied and taken to the ribosomes in the **cytoplasm** where the proteins are made.

How does the nucleus control cell division? First, by copying all the chromosomes and then dividing into two; this is followed by the cell cytoplasm dividing.

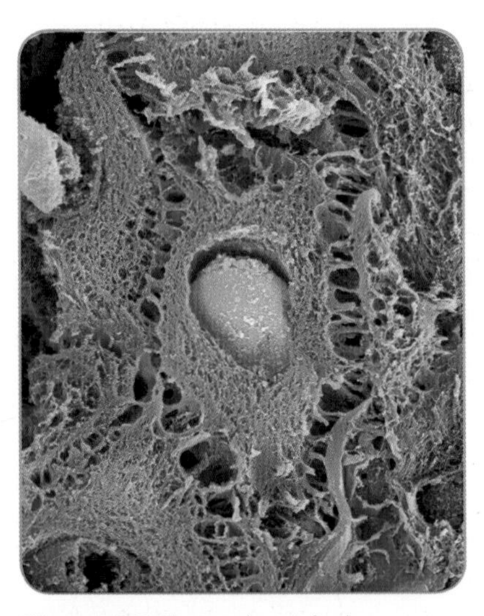

Figure 5.3 The nucleus of a human skin cell.

Cell membrane

The cell membrane surrounds the cytoplasm. The cell membrane controls the movement of many substances in and out of the cell. For example, the glucose used in the mitochondria for respiration enters the cell through the membrane, and insulin produced in pancreas cells leaves these cells through the membrane.

Cytoplasm

The cytoplasm is mostly water and fills the cell membrane. It is where most of the cell's chemical reactions take place. Proteins and enzymes are made in the cytoplasm. Enzymes control all the reactions that take place in the cell. The cytoplasm stores the dissolved raw materials, such as amino acids and glucose, needed for all cellular reactions. These reactions include respiration and protein synthesis.

Some of these enzymes pass out of the cell membrane to control reactions outside the cell. For example, amylase is an enzyme produced by cells in your salivary glands. Amylase passes out of the cells and digests starch into sugar in your mouth and oesophagus. Ideas about enzymes are developed in Chapter 7.

Mitochondria

Your whole body needs energy to keep warm. Cellular energy is used for all chemical reactions, to grow and for muscle activity. Mitochondria release the energy from glucose molecules during a process called respiration. Respiration is a series of chemical reactions that take place on the surface of the channels you can see inside the mitochondrion shown in Figure 5.4. These channels have a large surface area so that more respiration reactions can take place.

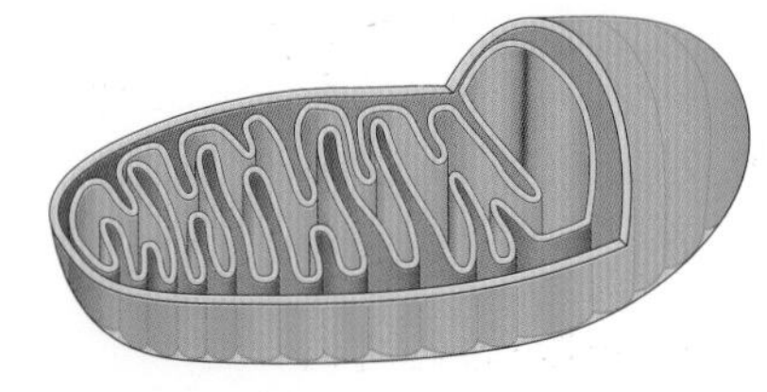

Figure 5.4 A mitochondrion.

Boost your grade

Some of the diagrams used here are three-dimensional to help you understand that the cell is a three-dimensional structure. You will only be expected to draw or complete two-dimensional diagrams in an examination (as Figure 5.6).

Test yourself

4 Muscles use a lot of energy for contraction. Which sub-cellular parts will you expect to see in large numbers in muscle cells?

5 Cells of the pancreas produce insulin, which is a protein. Which sub-cellular parts will you expect to see in large numbers in cells of the pancreas?

Ribosomes

The genes in the nucleus control production of **proteins** in the cytoplasm. Ribosomes are structures that link together the amino acids that make up proteins. The amino acids are linked in the order specified by the genetic code.

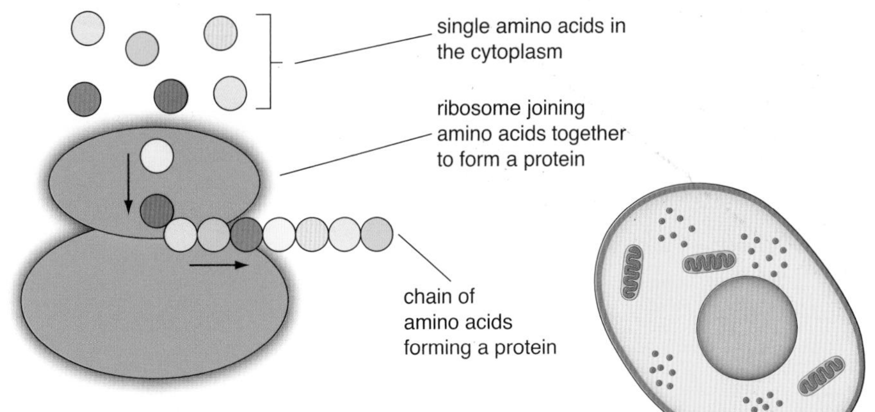

Figure 5.5 A ribosome forming an amino acid chain to be made into protein.

Figure 5.6 Animal cell.

Check Figure 5.6 and make sure that you can find and name all the parts.

Mitochondria are the parts of a cell responsible for releasing energy from glucose by respiration to make cellular energy (to talk about one use 'a mitochondrion' or 'the mitochondrion').

Ribosomes join amino acids in the correct order to make a specific protein. Each protein is coded for by a gene.

Proteins are long chains of **amino acids** that form the basis of many cell structures such as cell membranes, mitochondria and ribosomes, and body structures such as skin, muscle and hair. They also form chemicals in the body such as enzymes.

B2 5.2

Learning outcomes

Know and be able to describe the differences between animal, plant, bacterial and yeast cells.

Test yourself

6 Why don't animal cells need cell walls?

Chloroplasts use light energy to make glucose and oxygen from carbon dioxide and water by photosynthesis. They are found only in some plant cells but in most algal cells.

Plant, algal, bacterial and yeast cells have additional structures

Plant cells

What components can you find that are not in an animal cell?

Plant cells are larger than animal cells but we still need a microscope to examine them. Plant and algal cells have all the parts of an animal cell described in the last section and, in addition, they have the following.

Cell wall

Plant and algal cells have a rigid cell wall around the cell membrane that strengthens the cell and keeps its shape. It is made from a chemical compound called cellulose. Without cell walls plants would not be able to stay upright.

Chloroplasts

Chloroplasts are structures that contain a chemical called chlorophyll. Chlorophyll absorbs light energy to make food for the plant or alga(e) by a

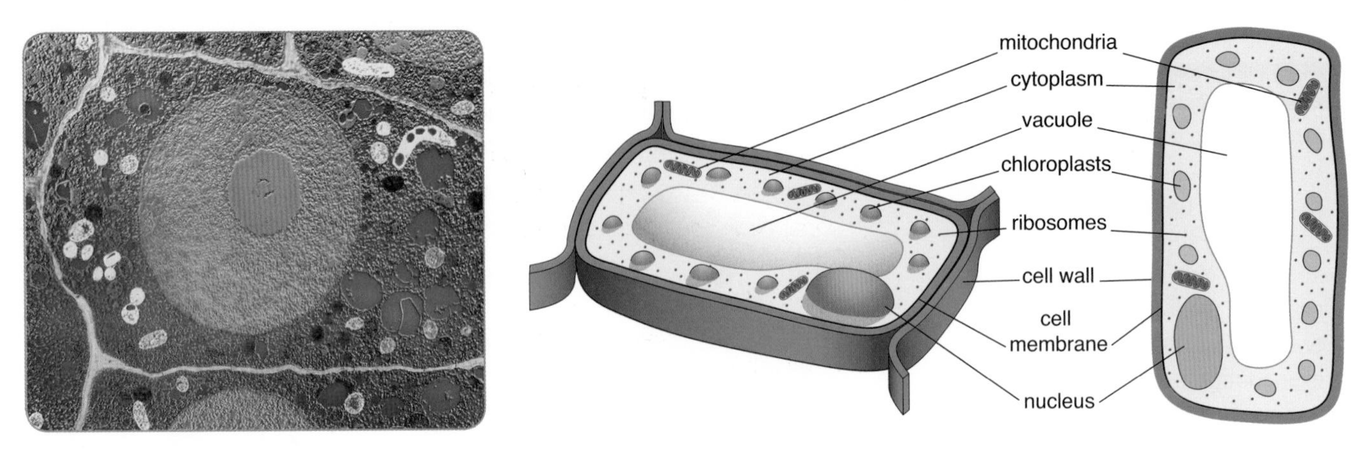

Figure 5.7 A plant cell shown as seen with a microscope (magnified ×2800) and as three- and two-dimensional diagrams.

process called photosynthesis. Photosynthesis is actually a series of chemical reactions that are all controlled by enzymes.

Permanent vacuole

The vacuole in plant and algal cells contains cell sap. Cell sap contains water, mineral ions and some dissolved chemicals such as sugars. The vacuole acts like a reservoir for the cell.

The vacuole presses against the cell wall and helps to support the cell by keeping it rigid.

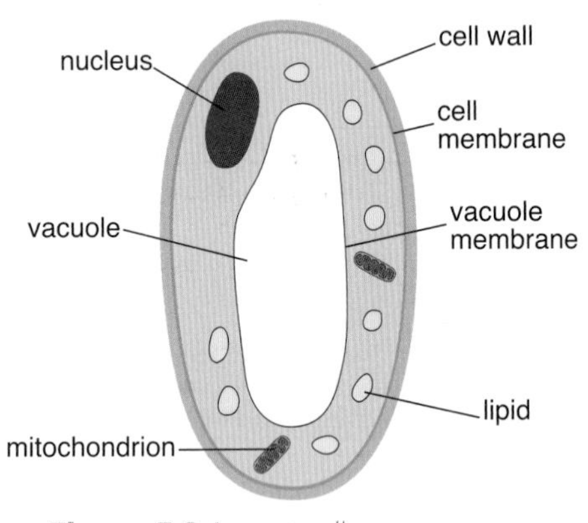

Figure 5.8 A yeast cell.

Yeast cells

Yeast is a single-celled fungus. It is important to humans because it can be used to make bread, beer and wine.

Yeast cells have a nucleus, cell membrane, cytoplasm, mitochondria and ribosomes with the same functions as plant and animal cells. A yeast cell also has:

- a fungal cell wall for support which is *not* made of cellulose, and
- a large vacuole containing dissolved substances such as mineral ions and sugar

Bacterial cells

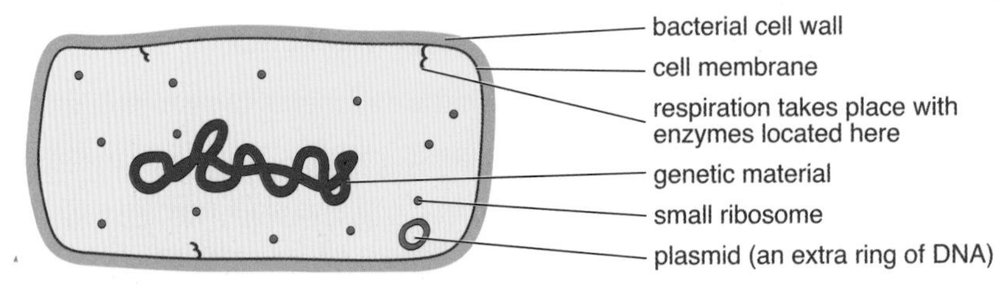

Figure 5.9 A bacterial cell (e.g. *E.coli*).

If you look at an evolutionary tree you will see that **bacteria** evolved before plants and animals. They do not have all the parts that plant and animal cells have.

Cytoplasm

The bacterial cell contains cytoplasm in which the chemical reactions of the cell take place. Small ribosomes are present in the cytoplasm for protein synthesis.

The word bacterium is singular: 'a bacterium'. We usually talk about many bacteria. The word can also be used as an adjective, for example, a bacterial cell.

Bacterial cell wall

This gives structural support to the cell. It is *not* made of cellulose.

- There is no nucleus.
- The genetic information of the cell is a long chromosome, which is free in the cytoplasm. The chromosome has the bacterial genes. (In the section on genetic engineering you saw a small ring of DNA called a plasmid — this is extra DNA.)
- There are no mitochondria.
- Respiration takes place using enzymes attached to extensions of the cell membrane.

Test yourself

7 Copy and complete this table

Table 5.1 Differences between plant, animal, yeast and bacterial cells.

	Plant and algal cell	Animal cell	Yeast cell	Bacterial cell
Nucleus	present			
Cell membrane				
Cell wall				Present — not made of cellulose
Mitochondria				
Chloroplasts				
Ribosomes				
Vacuole				

ICT

Create a resource that shows the similarities and differences between plant, animal, yeast and bacterial cells.

B2 5.3 How do cells perform specialised functions?

Learning outcomes

- Know the features and identify a range of specialised cells.
- Understand how to relate the structure of a specialised cell to its function.

Cell specialisation

We have seen the specialised functions in the sub-cellular parts and now we will look at how some cells have special shapes for particular functions. Remember that an embryo starts as a ball of cells that specialise as the organism develops. We say that the cells **differentiate** so that they can perform different functions. Think of specialised cells you may already know — red and white blood cells, nerve cells, the light receptor cells, specialised cells in the pancreas to produce insulin, a fibroblast cell making new connective tissue (see the photograph at the chapter opening). This process is essential for the body to heal after an injury. These cells are mature and, once specialised, do not change shape or function.

As cells mature they become different shapes or have different enzymes as they become specialised. We call this process differentiation.

However, there are some cells, called stem cells, which begin as being unspecialised but can produce a range of different cells. Stem cells are found in embryos, adult bone marrow (for forming the different blood cells) and some other body tissues such as in the liver and even at the bed of your fingernails.

Test yourself

8 Do specialised animal cells such as a muscle cell contain the same sub-cellular parts as skin cells?

A **tissue** is a group of cells with similar structure and function.

Epithelium means membrane. It can also be used as an adjective: epithelial cells.

Tissues

Large multicellular organisms need to group similar cells together to form body **tissue**, as the examples in Figures 5.10–5.12 show.

In these diagrams you can see that the cells within the tissue are the same shape.

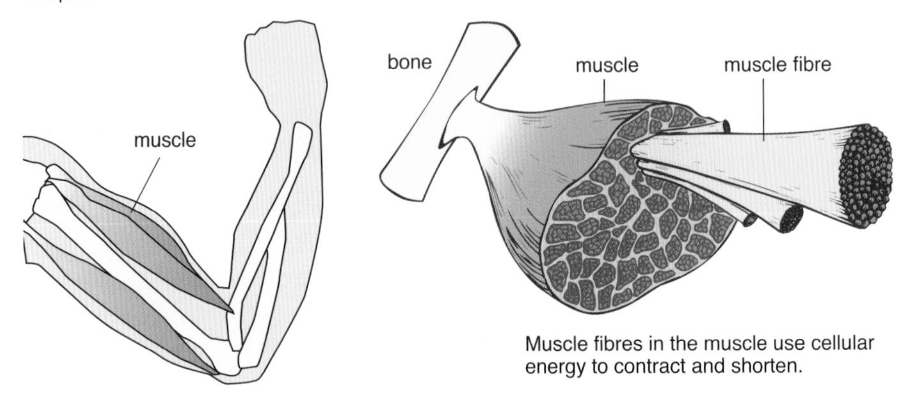

Figure 5.10 Muscle tissue is specialised to contract to bring about movement.

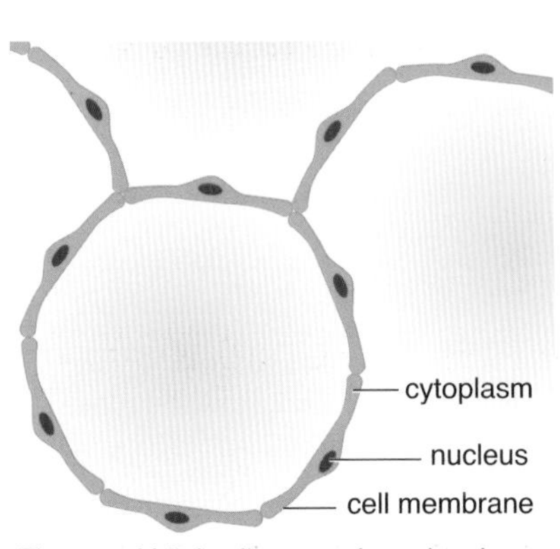

Figure 5.11 Epithelial tissue covers or lines some parts of the body. The thin **epithelium** lining the lungs allows gases to diffuse through.

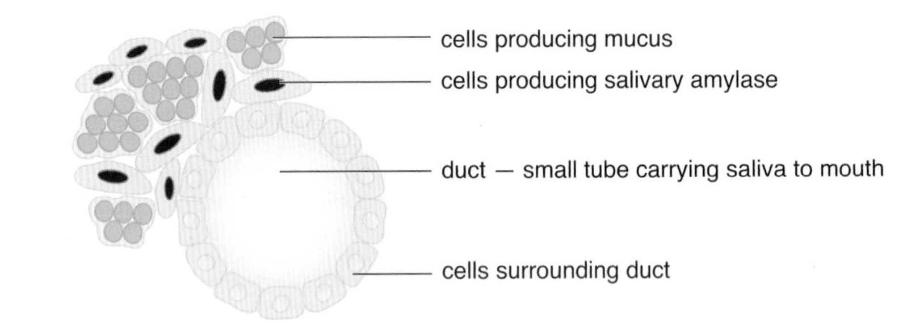

Figure 5.12 Glandular tissue produces substances such as enzymes and hormones.

B2 5.4

Even more organisation levels to form the whole organism

Learning outcomes

- Know why organisms require organ systems.
- Know the features and be able to identify a range of specialised organs in plants and animals.
- Understand the linking of organs into systems.

Organs

Further organisation comes from grouping tissues into organs. An organ contains more than one type of tissue so that the organ has one important function. Think of the body organs you know. The heart is an organ made up of largely muscular tissue to pump the blood but is covered and lined with epithelial tissue.

The stomach is an organ that contains the three tissue types shown in Figures 5.10–5.12.

- The stomach wall contains muscular tissue to churn and mix the contents with digestive juices.
- The juices contain enzymes produced by the glandular tissue.
- Epithelial tissue covers the outside of the stomach and lines the inside.

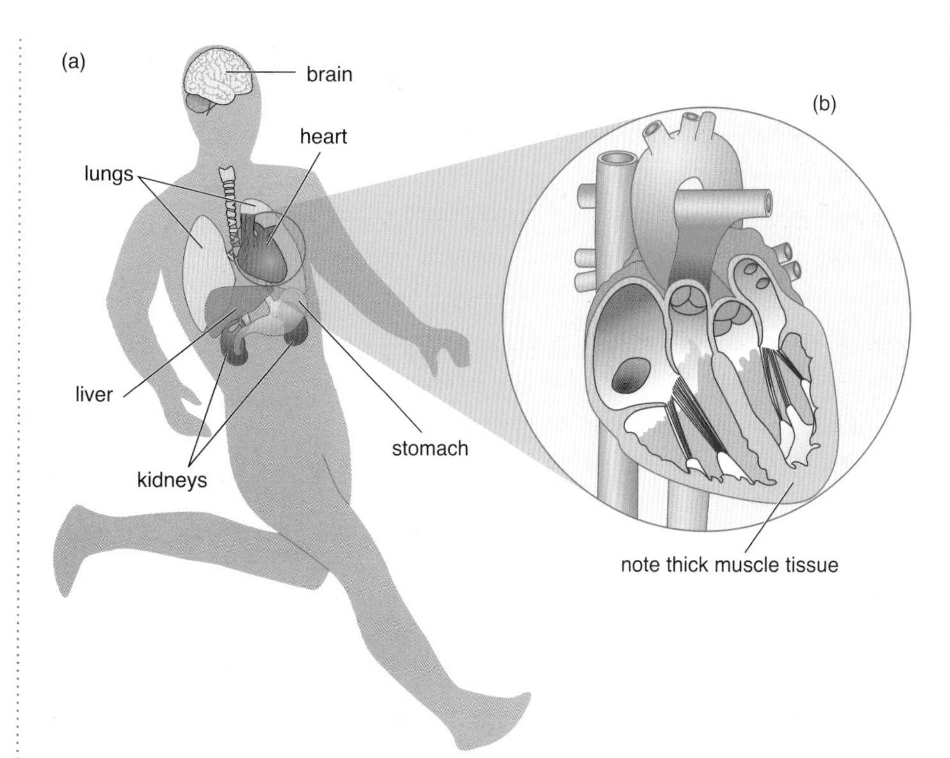

Figure 5.13 (a) Note the organs in this athlete, particularly the heart. (b) Note the thick muscles of the heart.

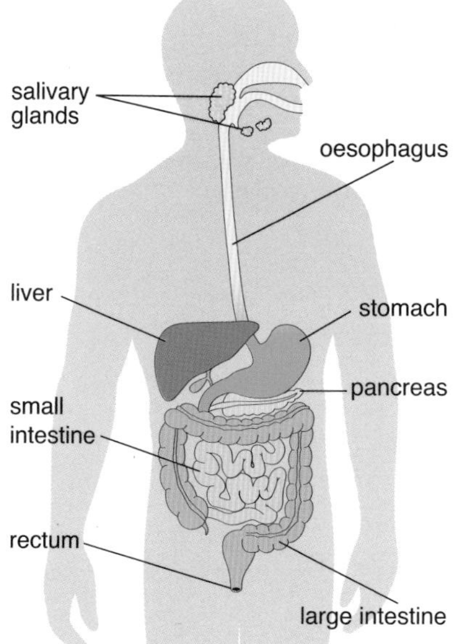

Figure 5.14 Digestive system.

Systems

To form the whole organism, the organs are grouped into systems so that one function follows another. Think about the digestive system:

- Food is taken in from the environment as mixed mouthfuls and has to be broken down chemically into small molecules that can be used by the body.
- Muscle tissue in the stomach wall mixes the food with the juices.
- Glandular tissue and organs produce the digestive juices containing enzymes that carry out these chemical reactions. The glands that produce the enzymes are the salivary glands, the stomach and the pancreas.
- The liver produces bile, which changes the pH in the small intestine to alkaline.
- The digestion takes place in the stomach and small intestine.
- Having produced small soluble molecules, such as amino acids and sugars, they must be absorbed into the blood further along the small intestine through the thin epithelial lining tissue of the villi (the villi are the tiny finger-like projections into the small intestine, which increase the surface area for absorption).
- This leaves a watery mass of indigestible matter. Water is absorbed in the large intestine and the indigestible matter is formed into faeces. This is stored temporarily in the rectum before being returned to the environment.

The whole organism is made up of all the different organ systems which function as a well-coordinated whole. As doctors know, if one system has a problem the whole body is affected.

Test yourself

9 List the following starting with the smallest:

mitochondrion, human chromosome, cheek cell, heart, plant cell, bacterial cell, digestive system

Plant tissues and organs

Multicellular plants have specialised tissues and organs. The roots anchor the plant and absorb nutrients and water, the stem is for transport and the leaf is an organ specialised for photosynthesis. The arrangement of tissues in the leaf is shown in Figure 5.15.

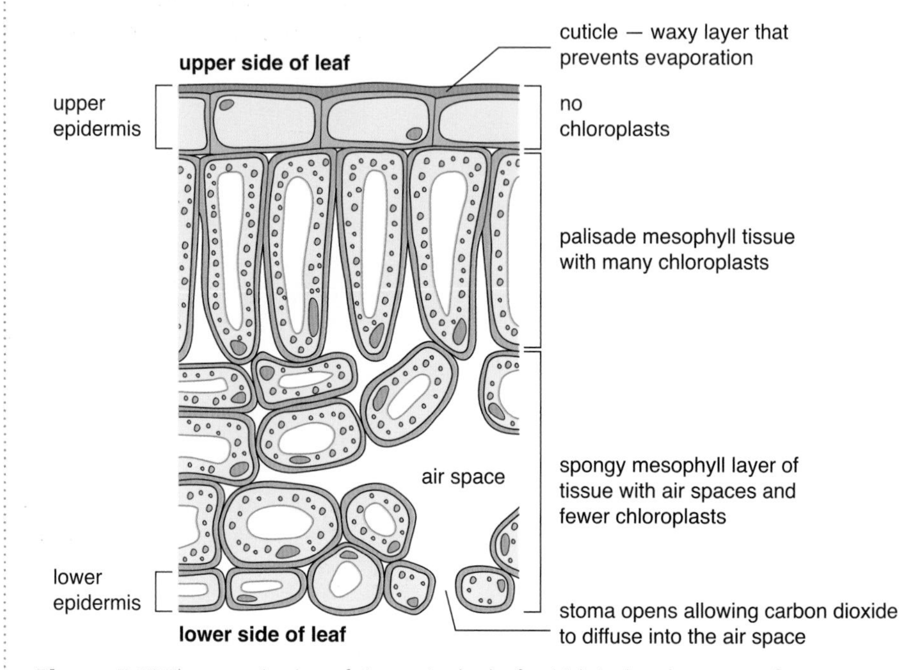

Figure 5.15 The organisation of tissues in the leaf, which is the plant organ for photosynthesis.

Test yourself

10 If the stem is the organ for transport in the plant, what two types of tissue will you find there?

11 Complete the statements and PRINT the correct level of organisation beside each statement. The first is completed for you.

The basic unit of all living things: CELL

Contains similar ________ cells performing the same function: ________

Made up of two or more ________ with related functions: ________

Structures with related functions are linked into: ________________

12 Where in a plant would you expect to find cells with the most chloroplasts? Give a reason for your answer.

13 Where in a plant cell would you find (a) DNA, (b) chlorophyll, (c) cellulose, (d) enzymes for respiration and (e) mineral ions in solution?

B2 5.5

Learning outcomes

- Explain how substances move into and out of cells by diffusion.
- Understand how this applies to examples in animals and plants.

How do dissolved substances move into and out of cells?

So far this chapter has referred to a range of processes, such as respiration and photosynthesis, which take place inside cells. These processes take place as a series of chemical reactions and therefore need chemicals that can get into a cell. They also produce chemicals that need to leave the cell. In this section we will consider how these substances move into or out of a cell.

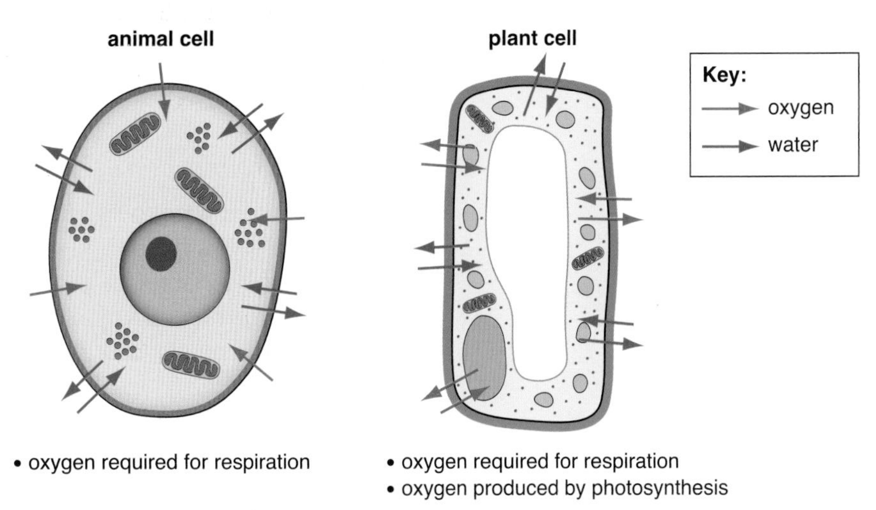

Figure 5.16 The movement of oxygen and water into and out of animal and plant cells.

Diffusion

If you spray air freshener in one corner of a room it is not long before it can be smelt throughout the room. This happens because the particles that make up the air freshener move around. To start with there are lots of particles in the space where you sprayed the air freshener, but quickly they spread out to where there are fewer particles. They move from a region where there is a higher concentration of particles to a region where there is a lower concentration. This process is called **diffusion**. Imagine if your class first huddled in the corner of the room and then ran about until they were evenly spread around the room. This gives a 'picture' or a model of how the particles behave when diffusion takes place.

Particles of a substance that is dissolved in a liquid will also diffuse. If you put a drop of ink in a beaker of water you can see the ink gradually spread throughout the water. This happens because the particles in the ink are diffusing amongst the water particles. They are moving from where there is a higher **concentration** of 'ink' particles.

How does this apply to living cells? Oxygen will dissolve in water and so can diffuse when in solution. This is how it moves into cells when it is used for respiration. Since oxygen is used up quickly in the mitochondria for respiration, there will always be a lower concentration of oxygen inside the cell than in the liquid outside the cell. We say there is a concentration gradient. This ensures that the oxygen will always diffuse into the cell. The greater the difference in concentration, the faster the rate of diffusion. Diffusion takes place because of the concentration gradient — it does not use energy.

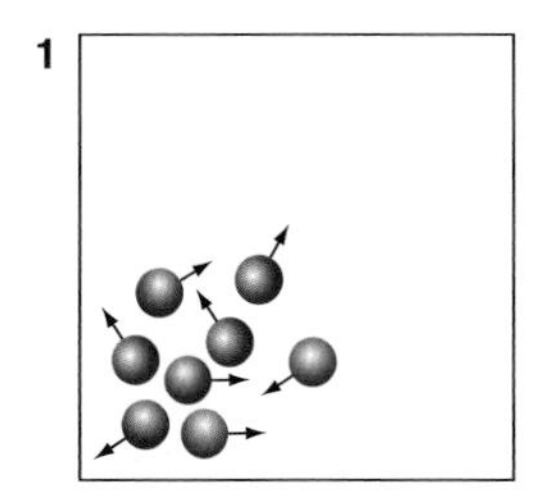

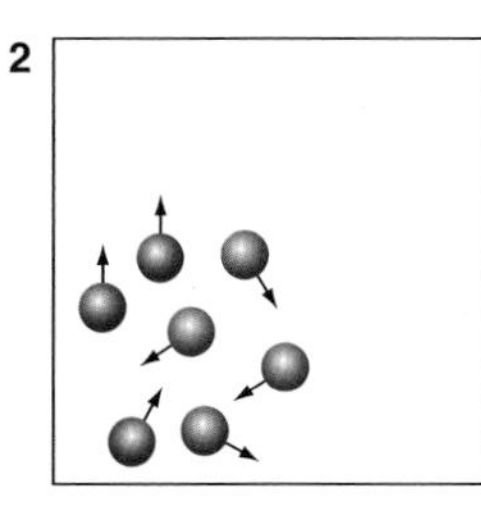

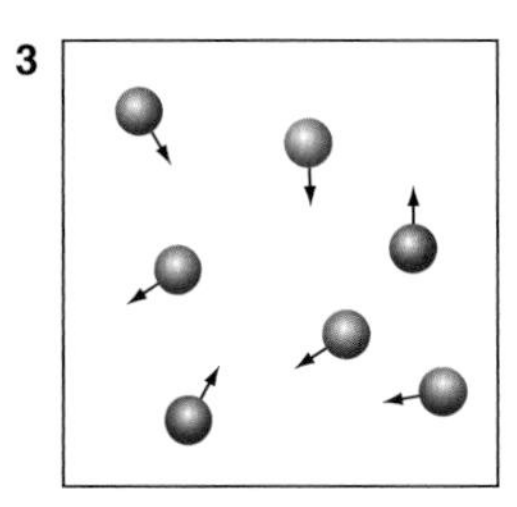

Figure 5.17 Molecules in a gas spread out by diffusion.

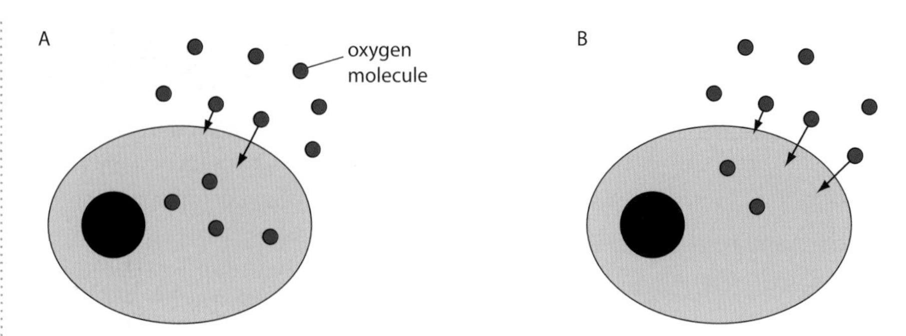

Figure 5.18 Count the number of molecules of oxygen in both cells. Into which cell will the oxygen diffuse faster?

Oxygen diffusion in the lungs

Blood transports oxygen to living cells but the blood first needs to be loaded with oxygen and this can only happen in the lungs. Diffusion will be faster if there is a large surface area for diffusion. In the lungs the surface area is increased by large numbers of alveoli. The concentration of oxygen in the lungs is kept high by breathing in air that contains 21% oxygen. The oxygen dissolves in the moisture of the lungs, diffuses through the thin epithelium and into the blood capillaries, diffuses into the red blood cells and is carried away as the blood is pumped along. Because the blood is always taking the oxygen away there is always a concentration gradient. Body cells produce carbon dioxide by respiration. This is carried back to the lungs in the blood. It diffuses in the opposite direction from the oxygen.

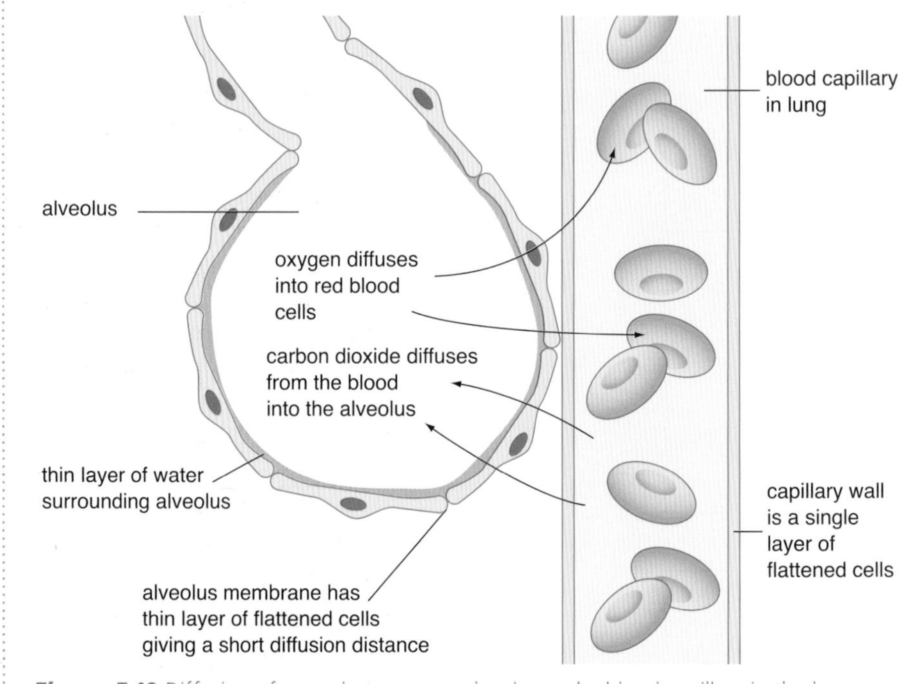

Figure 5.19 Diffusion of gases between an alveolus and a blood capillary in the lung.

Of course, oxygen, or anything else trying to get into a cell, has to pass through the cell membrane. The membrane will not let through larger molecules such as starch. Therefore the cell membrane is referred to as a **partially permeable membrane**. Small water-soluble molecules such as glucose can diffuse quickly.

Test yourself

14 What gas must diffuse into active muscle cells? What gas diffuses out of active muscle cells?

A simple demonstration of diffusion — ACTIVITY

Put a drop of strong-smelling oil into a balloon, blow the balloon up and tie it. Can you smell the oil from outside the balloon? If so, explain how it got from the inside of the balloon to the outside through the membrane of the balloon. How is this similar to what happens in the lungs?

Concentration is the amount of one substance in a given volume of another.

Diffusion is the movement of particles in a gas or any dissolved substance from a region of higher concentration to a region of lower concentration.

A **partially permeable membrane** is a membrane, such as a cell membrane, that will allow only some substances to pass through it.

Diffusion

HOW SCIENCE WORKS
PRACTICAL SKILLS

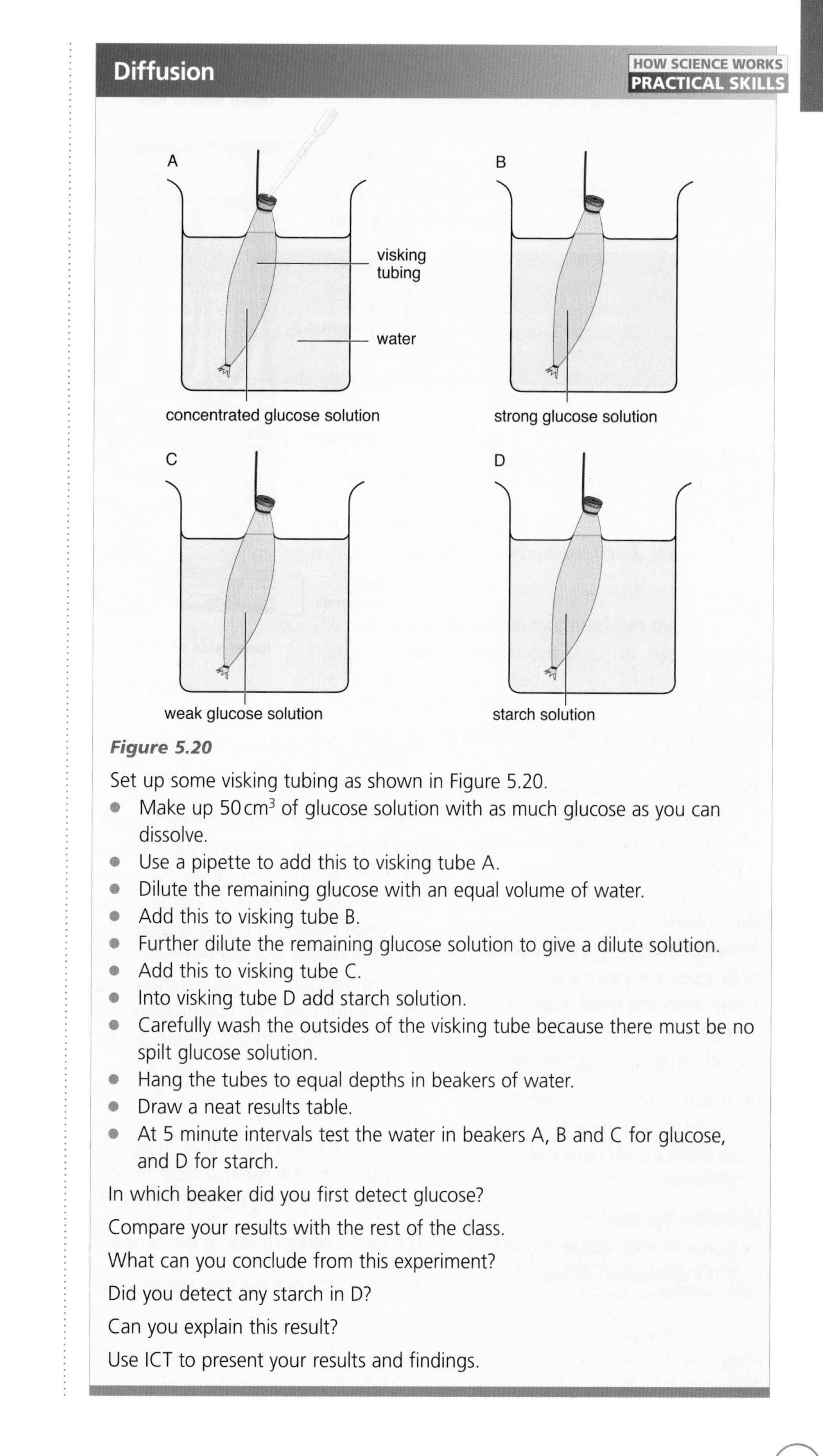

Figure 5.20

Set up some visking tubing as shown in Figure 5.20.

- Make up 50 cm^3 of glucose solution with as much glucose as you can dissolve.
- Use a pipette to add this to visking tube A.
- Dilute the remaining glucose with an equal volume of water.
- Add this to visking tube B.
- Further dilute the remaining glucose solution to give a dilute solution.
- Add this to visking tube C.
- Into visking tube D add starch solution.
- Carefully wash the outsides of the visking tube because there must be no spilt glucose solution.
- Hang the tubes to equal depths in beakers of water.
- Draw a neat results table.
- At 5 minute intervals test the water in beakers A, B and C for glucose, and D for starch.

In which beaker did you first detect glucose?

Compare your results with the rest of the class.

What can you conclude from this experiment?

Did you detect any starch in D?

Can you explain this result?

Use ICT to present your results and findings.

GET THE GRADE

Exam corner

1 A leaf is well adapted as the organ of photosynthesis. Explain this. *(4 marks)*

2 List the differences between plant cells and bacterial cells. *(3 marks)*

Student A

1 For photosynthesis a leaf needs to trap light energy. It has special cells called palisade cells which have lots of chloroplasts ✓ which capture the light energy. It needs carbon dioxide and this diffuses in through open stomata in the bottom of the leaf and between the spongy cells. ✓ It needs water and this is transported in the phloem. ✗ Because it is using the carbon dioxide to make sugar there is always a shortage of carbon dioxide so it always diffuses in from the higher concentration of the air outside. ✓

2 Cell wall, no mitochondria, no nucleus. ✗

Examiner comment You need to think through a question such as this before you start writing. It is a good idea to jot down the key words that you want to use.

Student A's answer to question **1** contains no reference to tissues — it only refers to cells. Did she miss the word 'organ' in the question? In question **2**, if you are asked a question about differences you must make it clear which cell you are referring to, i.e. 'A plant cell has … but a bacterial cell has …' Always use this format and you can't go wrong.

Student B

1 The leaf organ has layers of tissues. ✓ The top layer is the palisade mesophyll and the cells in this layer all have many chloroplasts to trap light. ✓ The bottom layer tissue is spongy so carbon dioxide can diffuse quickly ✓ through the spaces. The water is transported to the leaf in the xylem. ✓ The light energy is used to combine the carbon dioxide and water to make glucose. Carbon dioxide is used up so quickly that there is none in the leaf so it diffuses into the leaf quickly. ✓

2 The plant cell has a cellulose cell wall but the bacterial cell wall is not made of cellulose. ✓

The plant cell has mitochondria but there are none in bacteria. ✓

The plant cell has a nucleus but there is none in the bacteria — its chromosome is loose in the cytoplasm. ✓

Examiner comment Student B's answer to question **1** shows that she understands that an organ is composed of tissues and she explains this well. The point that the carbon dioxide is used up etc. is good but would have been stronger if it had referred to concentration differences or concentration gradient. The answer to question **2** is excellent — just what examiners are looking for.

Chapter 6 Photosynthesis, organisms and communities

Setting the scene

Why do you think these tomatoes are grown in a greenhouse rather than outside? What conditions might the commercial grower be trying to control inside the greenhouse?

ICT

In this chapter you can learn to:

- create a podcast for vegetable growers with advice on increasing yield
- use a social networking platform to discuss ideas with a wide range of students

Practical work

In this chapter you can learn to:

- make observations of cut flower stems and use them to explain how water travels through the plant
- identify which variables to control in a photosynthesis experiment and identify patterns in the results
- take measurements in a survey of land, calculate means and look for patterns in results
- evaluate the data from a grass survey
- work safely when investigating soil samples

B2 6.1 Photosynthesis

Learning outcomes

- Understand the process of photosynthesis.

Photosynthesis is the process in green plants which captures light energy and converts it to chemical energy used to make carbohydrates. The carbohydrates are used to form biomass. The word photosynthesis can be split into two parts: 'photo' meaning 'light' and 'synthesis' meaning 'to make'.

Test yourself

1 The box above shows that a condition for photosynthesis is light energy. What must be present in the leaf cells to trap this light energy?

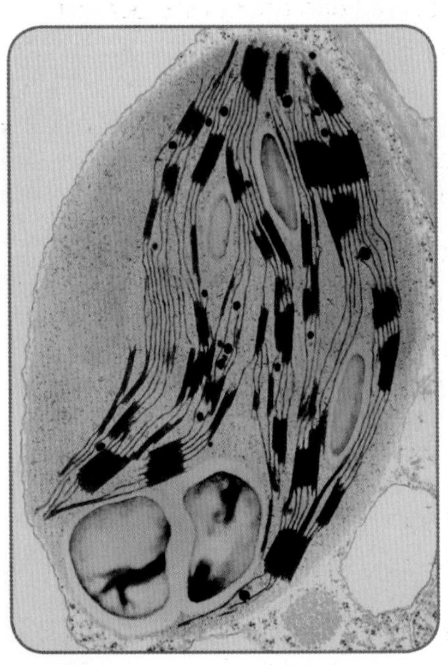

Figure 6.1 The structure of a chloroplast (magnification about ×15000).

Green plants use light to convert the simple molecules carbon dioxide and water into larger glucose molecules using enzymes in chloroplasts. This process is called **photosynthesis**.

We can break down the summary equation for photosynthesis into three parts, as shown below.

Raw materials	Conditions	Products
Carbon dioxide + water $6CO_2 + 6H_2O$	Light energy →	Glucose + oxygen $C_6H_{12}O_6 + 6O_2$

How is the light captured?

Some plant cells have chloroplasts to absorb light energy. Figure 6.1, a photograph taken at high magnification with an electron microscope, shows that the chloroplast has many membranes inside it. These membranes contain chlorophyll, which makes the leaves appear green. When light reaches the chlorophyll, enzymes in the chloroplasts use this energy to catalyse reactions that produce glucose. Glucose is quickly converted to starch that can be stored in the leaf, and oxygen is released as a waste product.

How is the raw material, carbon dioxide, obtained?

During the day, the leaf is photosynthesising and using carbon dioxide. The **stomata** (small pores) are open and the carbon dioxide diffuses into the leaf from the air. Diffusion always takes place from a region of higher concentration to one of lower concentration. Although the concentration of carbon dioxide in the air is on average only about 0.037% (or 370 ppm), when the leaf is

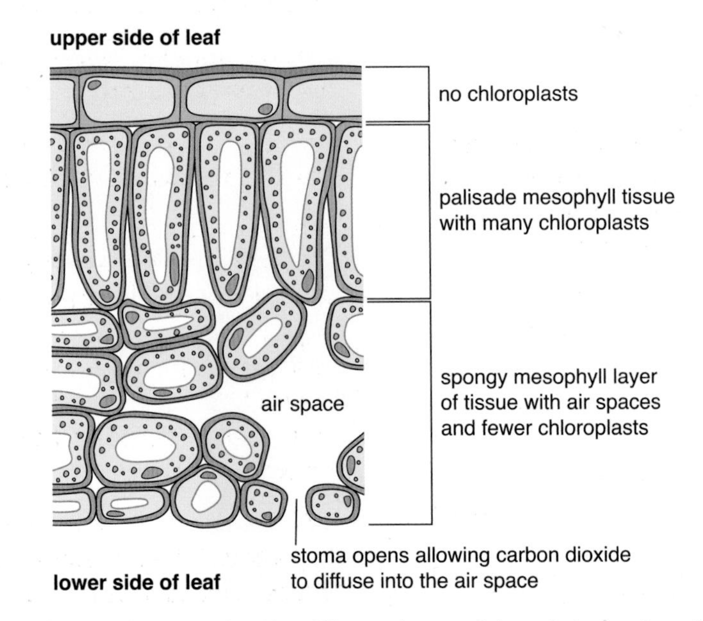

Figure 6.2 In this leaf section note the different shapes of the cells in the tissue layers.

photosynthesising it will be using up the carbon dioxide inside the leaf and so maintaining the concentration difference needed for diffusion.

The carbon dioxide concentration of the atmosphere can be written as a percentage or as ppm (parts per million). Atmospheric carbon dioxide levels are changing, and in 2006 a record high level of 381 ppm was recorded.

How is the raw material, water, obtained?

Roots take up water from the soil. The uptake is maximised because the root surface area is increased by many tiny root hairs. The water is then transported up the xylem vessels in the stem to the leaves. Oxygen is a valuable by-product from the reaction and diffuses out of the stomata into the atmosphere.

Visible water transport

PRACTICAL SKILLS

You can see which parts of the stem take water to different parts of the leaf by cutting a white carnation flower stem lengthways into several strips. Put each of these strips into a different coloured beaker of dye. This will create a 'tie-dyed' flower. Photograph your results.

Test yourself

2 What materials are needed for photosynthesis?

3 What is the energy input for photosynthesis?

4 A variegated plant has many white leaves.
   a What essential condition for photosynthesis is missing from these leaves?
   b Explain why this plant would not grow as rapidly as a green-leaved plant.

5 What are the products of photosynthesis?

B2 6.2

Learning outcomes

- Know the factors that affect the rate of photosynthesis.

What factors affect the rate of photosynthesis?

A reaction rate is the quantity of product produced or reactant used per unit time. You could compare this to a model with which you are familiar, such as the money you earn for washing cars on a Saturday job. For example, if you get paid £2 per wash and you wash two cars an hour, your hourly rate is £4.00 per hour. If you wash five cars an hour, your hourly rate is £10 per hour. The more cars you can wash in an hour, the greater your hourly rate of pay.

You can measure the rate of photosynthesis by finding the volume of oxygen produced per hour using a range of input values.

To find out how the input of carbon dioxide, light intensity and temperature limit the rate of photosynthesis, we need to measure the **rate of photosynthesis**.

From the equation for photosynthesis given in Section 6.1 you can see that carbon dioxide is one of the raw materials. Although leaf cells respire 24 hours a day, carbon dioxide cannot leave the leaf at night, since the stomata are closed (light intensity controls when the stomata open). By daybreak there will be a build-up of carbon dioxide in the leaf from the respiration that has taken place overnight. Initially light intensity will be a **limiting factor** for photosynthesis, but as the day gets brighter there will be more rapid photosynthesis, using up the carbon dioxide.

Water is essential for support of the leaf cells; if there is a shortage of water, the stomata will close to reduce water loss and photosynthesis will stop. This may happen around midday on a hot sunny day when there is a high water loss from the leaves or during drought conditions when the soil is dry.

Test yourself

6 Why does the concentration of carbon dioxide in the atmosphere limit the rate of photosynthesis?

7 What factor limits the rate of photosynthesis at dawn?

The **rate of photosynthesis** is the speed at which the photosynthesis reaction takes place. It can be measured as the rate of formation of oxygen.

A **limiting factor** is the factor that controls the rate of the product formation.

How does carbon dioxide concentration affect the rate of photosynthesis?

Carbon dioxide forms a small percentage of atmospheric gases and can be a limiting factor when the light intensity is high.

Analysing the effect of carbon dioxide concentration on the rate of photosynthesis

HOW SCIENCE WORKS
PRACTICAL SKILLS

A group of students decided to investigate the effect of carbon dioxide concentration on the rate of photosynthesis. They decided to calculate the rate of photosynthesis by measuring the volume of oxygen produced from an aquatic plant, *Elodea* (see Figure 6.3).

To investigate the effect of carbon dioxide concentration, the students made a **series dilution** of sodium hydrogencarbonate solution. They started with a 5.0% solution. They followed these instructions to make solutions with different concentrations.

- Take six 1000 cm^3 beakers. Label them A to F.
- Fill beaker A with 5% solution.
- Use a measuring cylinder to add 900 cm^3 of water into beaker B.
- Using a second measuring cylinder transfer 100 cm^3 sodium hydrogencarbonate solution from A to beaker B.
- Stir beaker B with a glass rod to mix. This gives a 0.5% sodium hydrogencarbonate solution.
- Make up a second beaker B as you require 2000 cm^3.
- Use the 0.5% solution to make up the solutions in the proportions shown in Table 6.1.

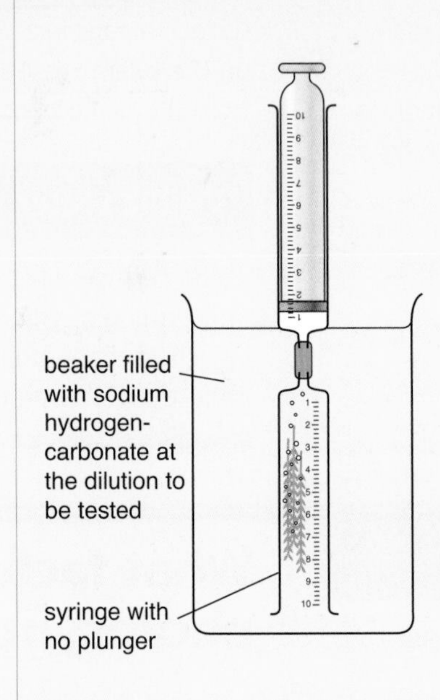

1 Connect the lower syringe to the empty upper syringe with a rubber connection.
2 Put three sprigs of *Elodea* in the lower syringe with the cut end upwards; there should be no air in this syringe.
3 Set in constant light for 10 minutes to equilibrate.
4 To start the experiment, draw any gas up into the top syringe and record the volume.
5 Leave the apparatus in constant light for at least 2 hours.
6 Draw up any gas produced in the lower syringe. Record the volume produced for each sodium hydrogen-carbonate concentration.

Figure 6.3 The apparatus and method used to measure rate of photosynthesis.

Table 6.1 A summary of the series dilution.

Beaker	Starting concentration of $NaHCO_3$	Volume of $NaHCO_3$ in cm^3	Volume of H_2O in cm^3	Final concentration
B	5%	100	900	0.5%
C	0.5%	800	200	
D	0.5%	600	400	
E	0.5%	400	600	
F	0.5%	200	800	

❶ Copy and complete the final column of Table 6.1.

The students used the apparatus and method shown in Figure 6.3. They knew that they should only change one input variable in each investigation. The input variable was the range of concentrations of sodium hydrogencarbonate solution. They knew that the light intensity and temperature should be kept constant.

❷ Which factor were the students controlling?

❸ This experiment was completed outside on a hot June day. What factors may not have been controlled?

To develop the experiment further the students decided to calculate the rate of photosynthesis in each experiment as the rate per gram of dry mass per hour. They measured the total dry mass of the *Elodea* used each time by placing it in a folded filter paper in a drying cupboard set at 100°C. The samples were dried and weighed every hour until the mass of dried *Elodea* remained constant for two successive weighings. Table 6.2 shows the set of results they obtained.

Table 6.2

Concentration of $NaHCO_3$ as %	Starting volume in cm^3	Final volume in cm^3	Volume of oxygen in cm^3	Mass of *Elodea* in g	Time in h	Volume of oxygen in $cm^3/g/h$
0.5	0.2	9.6	9.4	1.341	2	
0.4	0.4	8.6	8.2	1.399	2	
0.3	0.3	7.1	6.8	1.439	2	
0.2	0.5	5.9	5.4	1.510	2	
0.1	0.4	4.8	4.4	1.522	2	

4. Copy the table and complete the calculations. Use the following equation to complete the final column:

$$\frac{\text{volume of oxygen in cm}^3}{\text{mass of } Elodea \text{ in g}} \times \frac{1}{\text{time (h)}}$$

5. Use the results in the table to construct a graph with volume of oxygen in $cm^3/g/h$ on the y-axis and concentration of sodium hydrogencarbonate solution on the *x*-axis.
6. Suggest a suitable title for this graph, which links the output variable with the input variable.
7. Describe the pattern shown by your graph.
8. Suggest how you would modify this experiment to investigate the effect of light intensity on the rate of photosynthesis.

A **series dilution** is a range of concentrations. A minimum of five dilutions from concentrated to weak would give the input variable for the carbon dioxide concentration. The students planned the dilutions to give sensibly spaced points on the *x*-axis of their graph.

How does light intensity affect the rate of photosynthesis?

Light intensity is the amount of light (energy) per unit area of the surface. Light intensity decreases rapidly with increasing distance from the light source.

First, work through this simple model to make sure you understand what **light intensity** means. Switch on a torch, hold it close to a wall and shine the beam at the wall — you will get a small circle of light. As you move the torch further from the wall the circle of light gets larger and dimmer. The light energy given out by the torch is the same, but the light energy falling on each square centimetre of the wall (the light intensity) is less when the torch is further away.

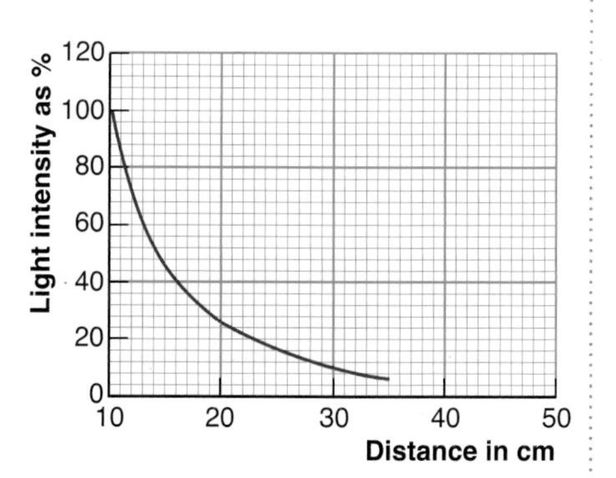

Figure 6.4 For this graph the light intensity was taken as 100% at 10 cm distance. Look at the shape of the graph. What happens to the light energy as the light source is moved away?

Light intensity is proportional to $1/d^2$, where d is the distance from the light source. This relationship shows that as the distance from a light source is increased, the light intensity reduces rapidly (Figure 6.4). You could use this graph of light intensity against distance to plan an experiment to find how light intensity affects the rate of photosynthesis. Unfortunately, normal light bulbs are inefficient and give off heat as well as light. Could this introduce another variable when the lamp is too close to the apparatus? Energy-efficient fluorescent bulbs or tubes give off less heat.

Light intensity generally increases towards midday and then decreases towards evening, but it can change quite quickly on a cloudy day. Light is the energy input and will have a significant effect on the rate of photosynthesis. The light energy capture stage of photosynthesis depends on the light intensity and is independent of temperature.

Test yourself

10 A plant was put into a black plastic bag and left for 24 hours. The leaves were then tested for starch. Explain why you would not expect to find starch in the leaves even during the day.

11 Where are the enzymes located that control the conversion of carbon dioxide to glucose?

12 When testing a leaf for starch it is dipped into boiling water to stop enzyme reactions.

a Suggest which enzyme reactions these are.

b Why does the alcohol become green?

c If parts of the leaf become blue-black when tested with iodine solution, it indicates that starch is present. What could you conclude if the iodine solution stayed brown?

13 Draw sketch graphs to compare the rate of oxygen produced on a bright winter day and on a bright summer day. Assume the light intensity is the same on both days.

14 A student grew a tomato plant in sand. After a month he compared it with his friend's that had been grown in potting compost. Both plants had been watered regularly but the plant in sand was shorter, had much smaller leaves and the lower leaves were yellow. Suggest reasons for these differences.

B2 6.4

Learning outcomes

Know the factors affecting photosynthesis in a range of habitats and agricultural situations.

What factors limit the rate of photosynthesis in the environment?

We have seen that under laboratory conditions light intensity, carbon dioxide concentration and temperature can all affect the rate of photosynthesis. In the environment all these factors interact, as you can see in the following activity.

From laboratory to outdoors

ACTIVITY

Use the information gained from laboratory examples to consider photosynthesis in the different environments shown below.

A field crop on a hot, sunny day (Figure 6.8)

1. Will light intensity be a limiting factor?
2. By 9 a.m. the rate of photosynthesis has reached a constant level and does not increase even though the light intensity is increasing. What could be the limiting factor?
3. At night the stomata are closed and the plant continues to respire, resulting in a build-up of carbon dioxide inside the leaf. What effect would you expect this to have on the rate of photosynthesis when it first gets light in the morning?

Figure 6.8

A forest in Iceland (Figure 6.9)

It has been said that you can't get lost in a forest in Iceland because the birch trees only grow to about 1.5 m high (see Figure 6.9). Iceland has periods of 24 hours darkness in winter and 24 hours light in summer, although the light intensity is not the same during those 24 hours.

4. What factors limit the growth of the trees?
5. Is carbon dioxide concentration likely to be a limiting factor?

Figure 6.9

A pond (Figure 6.10)

Ducks use the pond and their waste has resulted in extra nitrates (plant food) in the water. As a result there is dense weed growth and many snails feeding on the weeds.

6. What do you think will happen to the carbon dioxide concentration in the water overnight? Explain your answer.
7. During mid-morning, streams of tiny bubbles can be seen rising to the surface. What gas do you think these bubbles contain, and where does it come from?
8. What do you think happens to the carbon dioxide content of the water during the day? Explain your answer.

Figure 6.10

Figure 6.11 A large commercial greenhouse.

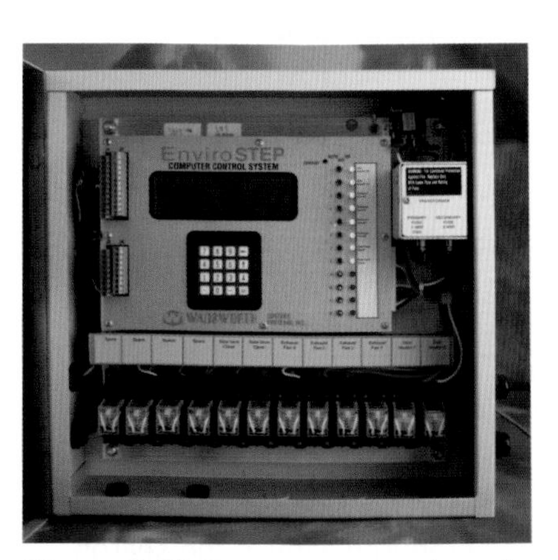

Figure 6.12 A computer control panel that controls conditions in a greenhouse.

Controlling the greenhouse environment

The environment in commercial greenhouses is closely monitored using sensors that measure temperature, carbon dioxide concentration and light intensity. If any of these factors change, computer controlled systems artificially restore the factor to optimum so that the plants receive ideal conditions of light, temperature and carbon dioxide. For example if a drop in carbon dioxide is detected a valve automatically opens to increase the concentration and closes again when the optimum concentration is reached. This saves adding it all the time and wasting the gas. Lights will switch on until the sun rises and the light intensity is high enough and then switch off until late afternoon when the light intensity drops.

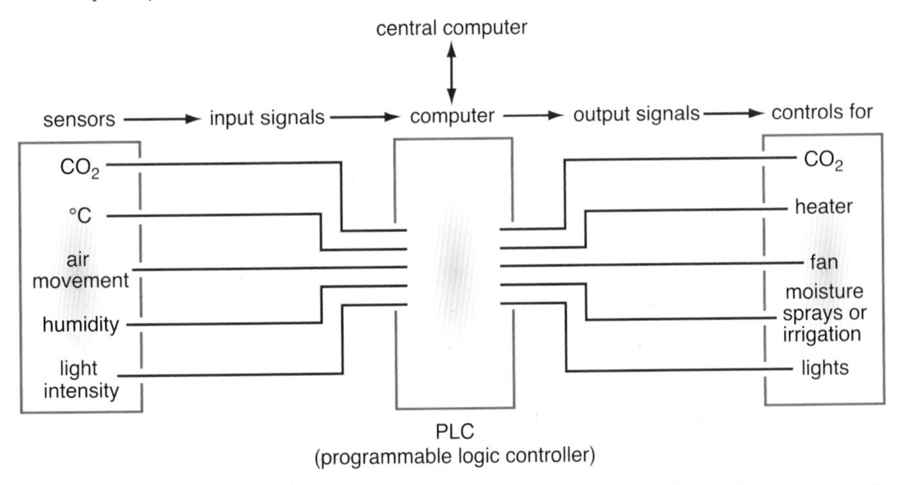

Figure 6.13 A control system for the greenhouse. Does this resemble any human system?

Chapter 7
Enzymes: function, industrial uses and their role in exercise

ICT

In this chapter you can learn to:

- create an animation of the lock and key model for how enzymes work
- use audio and visual resources to show the many uses of enzymes in industry

Setting the scene

Why is biological washing powder called biological when it is full of chemicals?

Practical work

In this chapter you can learn to:

- plan and carry out an investigation into how different conditions affect enzyme reactions
- use graphs to display the results of an investigation into the effect of exercise

B2 7.1

Learning outcomes

- Know that proteins act as structural components of muscles, hormones, antibodies and enzymes.
- Know that enzymes are proteins which act as biological catalysts that increase the rate of reactions in living organisms.
- Understand that protein molecules and enzymes are made of chains of amino acids coiled together.

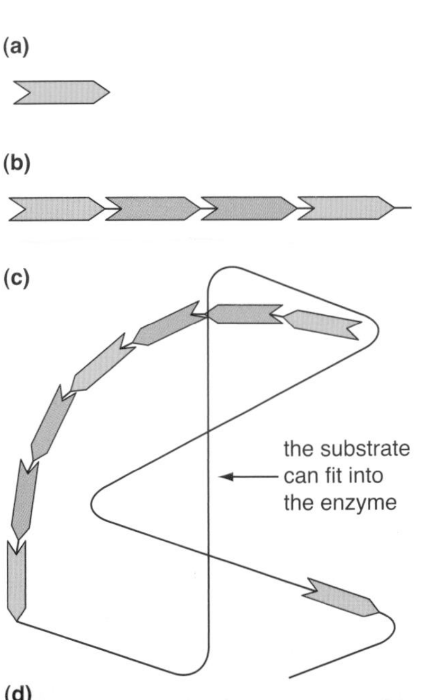

Figure 7.1 (a) A single amino acid; (b) a long chain of amino acids joined together; (c) for three-dimensional proteins the chain folds and forms bonds across the chain between amino acids to make it more stable; (d) a three-dimensional model of an enzyme molecule with the substrate in position.

Proteins are specialised biological molecules

Ribosomes assemble amino acids to form proteins. The order of assembly is dictated by the DNA code in the nucleus. Proteins have many different functions.

Building protein for growth

Growing cells need to build new structures inside the cell. To do this they make proteins. Proteins are components of cell membranes and the membranes of the sub-cellular parts such as mitochondria. Amino acids can be joined together to form structural proteins and enzymes. Enzymes catalyse the reactions that make proteins inside the cell. This is called protein synthesis — different enzymes control which proteins can be built. Different cells have different functions and so there are thousands of different proteins.

A protein that you will know well is keratin, which makes up hair, feathers and also hooves. It is a structural protein. Keratin is not soluble. Muscle tissue is composed of a protein that is specialised to contract.

Hormones are soluble proteins that are transported in the blood plasma to target organs. Think back to genetic engineering and the manufacture of growth hormone. The growth hormone molecule has about 200 amino acids. The chain of amino acids is folded into a specific three-dimensional shape so that the hormone can bind onto receptor sites of the target organ membranes.

Antibodies also have a surface shape. Antibodies are specific to the antigen of a pathogen. This means that the antibodies can attach to the pathogen surface and inactivate it.

Enzymes are biological catalysts. They are also three-dimensional molecules. Here we look at them in greater detail.

Test yourself

1. Name two structural and three soluble globular (three-dimensional) proteins.
2. What gives three-dimensional proteins their specific shape?

What are enzymes and what are their functions?

Enzymes are called biological catalysts because they control most of the reactions that take place in living organisms. All catalysts change the rate of chemical reactions but are not themselves changed by the reaction. Enzymes catalyse essential reactions such as respiration and photosynthesis that take place inside cells. The enzyme is not used up so it can catalyse the same reaction many times.

The function of enzymes is to enable reactions to take place at lower temperatures. Without enzymes, at low temperatures, such as the temperature of the human body, the chemical reactions inside cells would take place slowly.

In the food and biotechnology industries enzymes are used in many processes so that the reaction temperature can be kept low. This saves energy used to heat the container and so saves money. Enzymes also increase the reaction rate so that large amounts of product can be made quickly.

Enzymes are biological molecules that catalyse reactions in living organisms. Enzymes are referred to as biological catalysts.

Enzymes differ from chemical catalysts because they are protein molecules. Therefore, they have some properties of both catalysts and proteins. As catalysts, enzymes change the rate of reactions and as proteins they have a complex three-dimensional structure that allows other molecules to fit into the enzyme.

Test yourself

3 What is the benefit of enzyme-controlled reactions inside living cells?

4 Give three advantages of using enzymes in biotechnology applications.

5 Give two similarities and two differences between enzymes and other catalysts.

B2 7.2

How do enzymes work?

Learning outcomes

- Know that enzymes have an optimum temperature and optimum pH.
- Explain why the shape of an enzyme is vital for its function.
- Explain why heat affects enzyme function.

During digestion or breakdown reactions, each enzyme works on one particular substrate giving certain products. Different enzymes work on different food types. Protease enzymes speed up digestion of protein and amylase breaks down the substrate amylose (commonly called starch) to give glucose.

Why does an enzyme only work on one substrate? Let us look at the structure of enzymes in more detail. You will need to think about the three-dimensional structure of the molecule. Remember that enzymes are protein molecules.

Amino acids are the building blocks of proteins. They are joined end to end to form a long chain. This molecular chain is stabilised when it is folded into a three-dimensional shape and cross-bonds form between the amino acids (rather as a girl might twist up her hair and put in slides). All enzymes have this same basic structure, but they have a different sequence of amino acids to form different protein molecules. The order in which amino acids are joined together is controlled by the DNA code in the nucleus.

Each tightly folded protein molecule or enzyme has ridges and grooves, giving it a unique shape. The idea of the substrate fitting into the enzyme is called 'the **lock and key**' model for enzyme action. Just as two keys can look similar but will not work in the same lock, so similar substrates cannot be broken down by the same enzyme. This is why enzymes are described as specific. Once the correct substrate is attached to the enzyme molecule, a reaction at the surface active site (ridge or groove) of the enzyme takes place and a chemical bond is broken in the substrate.

The sequence of a *breakdown* reaction is:

- the enzyme and substrate collide
- the substrate attaches to a special ridge or groove on the enzyme surface
- the enzyme breaks the bond in the substrate
- products are released

Test yourself

6 Work out the sequence of the reaction to build up a single product from two substrate molecules — for example, to build up a molecule such as sucrose. The structure of sucrose is shown in Figure 7.2.

A **model** is a simplified picture that scientists construct to fit observed data to a theory.

The **'lock and key' model** suggests that each enzyme (the lock) has a special shape, which means that only one substrate (the key) can fit into each enzyme.

The **activation energy** is the energy that must be provided to the reactants to start a chemical reaction.

A **substrate** is the substance acted on by an enzyme.

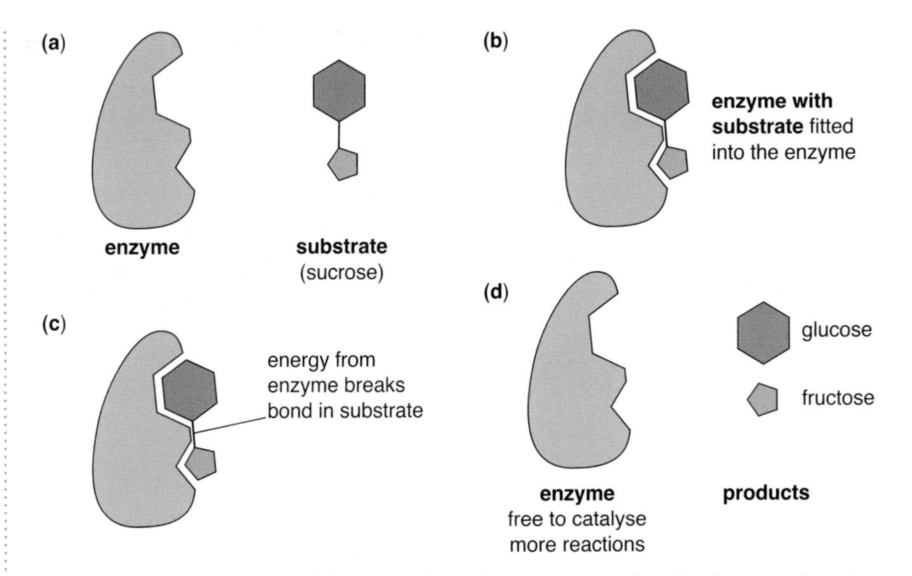

Figure 7.2 This sequence of diagrams shows how an enzyme breaks down a substrate. Here, the substrate is sucrose and the products are glucose and fructose.

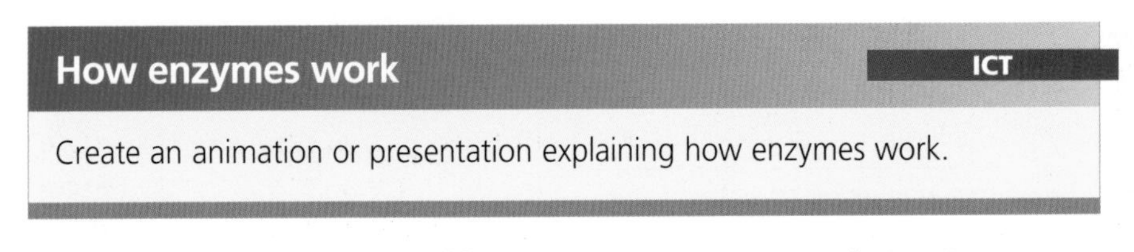

What conditions affect enzyme reactions?

Rates of enzyme reactions increase with increase in temperature. To understand why, you can apply your understanding of what is happening at the molecular level. First, the enzyme and substrate molecules must collide with enough energy for the substrate to fit into the enzyme structure and start the reaction. This is called the **activation energy**. This is a bit like putting a missing piece into a jigsaw puzzle: you can place a missing piece over a gap, but it needs a little push (energy) to fit it in.

An increase in temperature will cause the particles to move faster, that is, to increase their kinetic energy. As a result enzyme molecules and substrate molecules will collide more often and with more energy. The enzyme can catalyse more reactions so the reaction rate is increased.

But this is where enzymes differ from catalysts: enzymes function best at an optimum temperature (see Figure 7.3). An increase in temperature above this changes the shape of the protein molecule (see Figure 7.4). You can see the effect of change in the protein structure when you cook an egg white. When the enzyme molecule shape has been changed, the substrate can no longer fit into the enzyme — we say that the enzyme has been **denatured** and it can no longer function as a catalyst.

Living organisms are adapted to exist in a range of habitats, so the enzymes inside bacteria which live in hot springs will be able to function within a different temperature range from those enzymes in plants in the Arctic. There are also differences in pH levels at which enzymes work. Each enzyme has an optimum pH, which is the pH at which it functions best or most effectively.

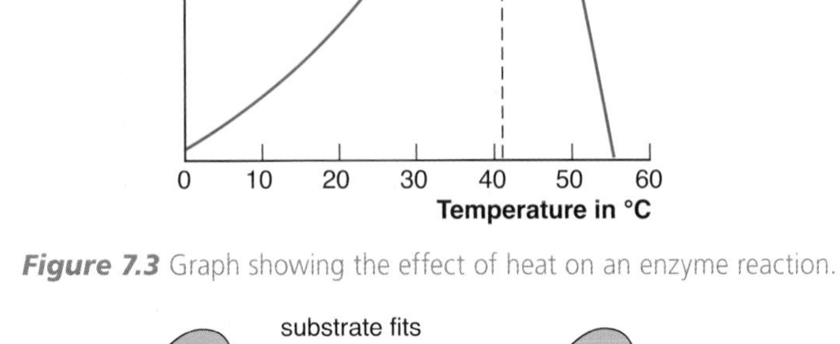

Figure 7.3 Graph showing the effect of heat on an enzyme reaction.

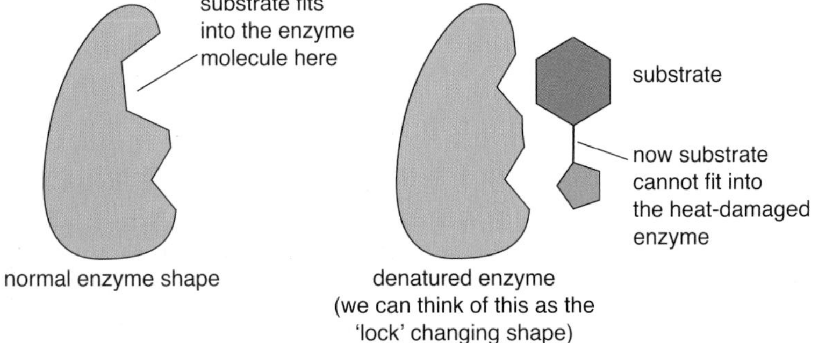

Figure 7.4 The effect of heat (denaturation) on an enzyme molecule. Heat alters the bonds and changes the three-dimensional shape.

When temperature destroys the three-dimensional shape of a protein molecule or enzyme, it is **denatured**.

Effect of temperature and pH

HOW SCIENCE WORKS
PRACTICAL SKILLS

Plan and carry out an experiment to show how one of these variables (temperature or pH) affects the action of enzymes.

Test yourself

7 Why are enzyme reactions slow at low temperatures?

8 Why does the rate of product formation increase as the temperature increases between 5°C and 30°C in an enzyme reaction?

9 A student was carrying out an experiment using the enzyme protease on boiled egg white. He found that the white solution became colourless for all temperatures up to 45°C, but above this temperature the solution remained cloudy. Can you explain this?

10 What term is given to the temperature at which enzymes function most efficiently?

11 The mouth is slightly alkaline but the stomach has a low pH. Why do you think salivary amylase stops working when the food it is mixed with enters the stomach?

12 Why can protease break down protein but not starch?

Digestion in action

ACTIVITY

How could you act out as a drama the process of digestion from a food being eaten and broken down to being absorbed in the small intestine and passing into the blood?

Who was Emil Fischer?

HOW SCIENCE WORKS ACTIVITY

Emil Fischer was born in Germany in 1852. His father, who said that Emil was too stupid to be a businessman and had better be a student, sent Emil to the University of Bonn in 1871 to study chemistry. Emil Fischer won a Nobel prize in 1902. This was the first prize awarded for organic chemistry. He was famous for rigorous teaching and making links between science and industry. Six of his students later went on to become Nobel prize winners.

Emil Fischer first suggested the 'lock and key' model in 1894. The amazing thing about Fischer's model was that he suggested this before scientists had discovered that enzymes were protein molecules.

Figure 7.5 Emil Fischer.

The 'lock and key' model is used today to develop computer simulations that investigate possible new drugs. The computer programs design and test thousands of different molecules with different shapes, to see how they could 'fix' onto receptors on the surface of target disease cells.

1. Write and illustrate a newspaper report about Fischer's great award. Find out about him by searching for his name and biography. How long after his discoveries was he rewarded?
2. Find the answers to the following questions using books or an internet search.
   a In Emil Fischer's model, what does the lock represent?
   b After the 'lock and key' model was suggested, how long was it before the nature of enzymes was worked out, and how long before a three-dimensional model of a protein was first made?
3. Draw a diagram to show the 'lock and key' model as a lock and key, and a more modern diagram showing an enzyme and substrate.

Test yourself

13 a The food leaving the mouth is slightly alkaline. How is the pH lowered in the stomach?
b What is the optimum pH for the enzyme amylase?
c The food leaving the stomach is strongly acidic. How is the pH raised?
d What is the optimum pH for all the pancreatic enzymes?
e What is the optimum pH for stomach protease?

14 At the start of the chapter, in question 3, 'What is the benefit of enzyme-controlled reactions inside living cells?' you considered temperature. Now think about the rate of energy production and give a second answer. Suggest how slow uncatalysed reactions would affect life.

15 Copy Table 7.2 and complete the four blanks.

Table 7.2

Enzyme	Large complex molecules	Small soluble end products of digestion
Amylase	Starch	(a) ______
Protease	(b) ______	Amino acids
(c) ___	Lipids (fats and oils)	(d) ______ and glycerol

16 a Name the two places where amylase is produced.
b Name the two sites of protein digestion.
c Describe where fats and oils are digested.

17 a Where is bile produced?
b Where is bile stored?
c What is the function of bile?

B2 7.4

Learning outcomes

- Understand and describe how enzymes produced by microorganisms are used in detergents to remove food stains.
- Understand and describe applications of enzymes used in industry including pre-digested baby food and products containing sugar or fructose syrup.
- Evaluate the advantages and disadvantages of using enzymes in the home and in industry.

Industrial production and use of enzymes

The list of industrial applications for enzymes is amazing and includes food, wine and beer production, making fructose sweeteners for soft drinks, laundry detergents, producing the 'stone-washed' effect in jeans and manufacturing pharmaceuticals. The food industry uses many enzymes, which can be 'tailor-made' to give an exact product for the development of a new food.

Like all living organisms, bacteria and fungi produce and secrete enzymes. The enzymes pass out of their cells into the environment. The enzyme products, small soluble molecules, are then taken into the organism's cytoplasm. Microorganisms are collected from different locations around the world and the enzymes they produce are tested. Biotechnologists look for enzymes that could be used for particular processes, and that work at the temperature of the industrial plant.

As bacteria and fungi normally live in cool environments, their enzymes usually function at low temperatures. However, if an enzyme is required in industry to work at high temperature or extremes of pH, then enzymes have to be found from microorganisms living in those environmental conditions. Enzymes that act on the right substrates but at the wrong temperature may need to be modified by gene transfer.

To produce the enzymes, the microorganisms are cultured in a fermentation process. At the end of the batch process, the **fermenter** is emptied and the enzymes are collected from the solution in which the bacteria have grown.

Many industrial enzymes come from soil microorganisms. One of the best sources of industrial proteases used in laundry detergents is *Bacillus*, a common soil bacterium.

A **fermenter** is a large steel vessel used for biochemical reactions. Sensors monitor the conditions inside. The sensors send information to a computer, which then controls input valves to maintain the temperature, pH, nutrient and oxygen levels at the optimum values.

Enzymes in the food industry

An early application of enzymes in the food industry was the production of sweet syrups by breaking down starch. This process can be done chemically by boiling starch with acid, but the reactions using the enzyme method give a pure and reliable product, with no other by-products. Another advantage is that energy costs are lower as the enzyme process can be carried out at a lower temperature.

Glucose syrup is widely used as an ingredient and food additive. Most glucose syrup is produced from maize (corn) using enzymes. Maize processing results in a large amount of starch waste products. Using a common and cheap (waste) material as the raw material for glucose syrup is cost effective.

Figure 7.8 An industrial fermenter for producing enzymes. The raw material is converted by the enzymes as it flows slowly over beads with enzymes trapped on the surface (see Figure 7.9).

Many **carbohydrase** enzymes are used in sequence to produce a variety of syrups of different sweetness. For example:

- maize starch is treated with one form of amylase to convert it to a thick starch paste
- this paste is then reacted with different amylase enzymes to form sugars such as maltose or glucose. The reaction can be stopped here, or
- the final stage is the conversion of glucose syrup to fructose syrup, using the enzyme **isomerase**

In the early 1970s a continuous flow system (see Figure 7.9) was developed in which glucose solution was added through the top and fructose solution was delivered from the bottom. The continuous process can run for about 6 months non-stop. As enzymes are proteins, their structure eventually breaks down and the enzyme needs replacing. The old enzymes are biodegradable in the environment so do not result in toxic waste. Fructose syrup is much sweeter than sucrose or glucose syrup so it can be used in smaller quantities as a sweetener. This saves food manufacturers money, but another result was that in the 1970s fructose syrup began to be used in slimming foods to give sweetness with fewer calories.

Although there are clear economic advantages of using enzymes in industrial processes, there are some disadvantages:

- development and production of new enzymes is costly
- enzyme action is highly specific, so they have only one use
- enzymes require exact conditions of temperature and pH in which to work
- as enzymes are proteins, they can cause allergic reactions, and so must be encapsulated to minimise the risk of skin contact or inhalation
- some people may object to genetically modified (GM) enzymes being used to produce food

Carbohydrase enzymes break down complex carbohydrate molecules such as starch into simple sugars.

Isomerase is an enzyme that converts glucose into fructose.

Test yourself

18 List the economic advantages of using enzymes in the food industry.

19 a In a continuous flow column, what is the advantage of having the enzymes immobilised on beads? (See Figure 7.9.)

b In a batch process, the raw materials and the enzymes are reacted together in a reactor vessel until an economic concentration of products has been formed. The remaining raw materials and the enzymes are then separated from the products. What are the advantages of 'continuous flow' rather than 'batch' processes?

to higher levels of fitness. Their results for this part of the investigation are shown in Table 7.6.

5 Plot both students' results as line graphs on one set of axes.

6 a What was the recovery time for each student?
b Which student does this suggest is the fitter?
c What other factors should be considered before drawing a firm conclusion from these data?

Table 7.5 Sara and Kerry's results.

	Sara		Kerry	
	Heart rate in beats per minute	Breathing rate in breaths per minute	Heart rate in beats per minute	Breathing rate in breaths per minute
Resting	68	12	72	12
Walking	63	12	86	12
Running on the flat	112	16	120	14
Running up hill	126	19	134	17

Table 7.6 Sara and Kerry's results for recovery time.

Time in minutes	Sara	Kerry
0	126	134
1	115	128
2	107	121
3	95	112
4	83	105
5	74	96
6	68	88
7	68	80
8	68	72
9	68	72

B2 7.8

Learning outcomes

- Explain how muscles respond to long periods of vigorous exercise.
- Explain how the body 'repays' an oxygen debt.
- Understand how an oxygen debt builds up in muscles during anaerobic respiration owing to the formation of lactic acid.
- Understand how to interpret data about the effects of exercise on the body.

Muscle fatigue occurs during long periods of vigorous exercise and causes the muscles to work less efficiently.

What are the effects of vigorous exercise?

Our muscles can keep working provided they have a sufficient supply of oxygen and glucose. Even when the supply of glucose is limited, our muscles convert glycogen into glucose in order to maintain respiration. However, during long periods of vigorous exercise this situation can change.

The person in Figure 7.15 is exercising hard. If the exercise continues for a long time his muscles will start to suffer from a condition called **muscle fatigue**. This occurs when the blood cannot supply the muscles with enough oxygen. When this happens, the muscles start to respire anaerobically.

Lactic acid and oxygen debt

During anaerobic respiration, energy is released from glucose without oxygen. However, anaerobic respiration releases far less energy per gram of glucose than aerobic respiration. This is because anaerobic respiration only partly breaks down glucose, producing a chemical called **lactic acid**. As lactic acid builds up in our muscles, it prevents them contracting properly. This is when muscle fatigue sets in. Just hold your arms straight out in front of you for a long time if you want to experience muscle fatigue!

Figure 7.15 Exercising vigorously for long periods of time can put your muscles under great stress.

Lactic acid builds up in muscles during anaerobic respiration.

An **oxygen debt** is the amount of oxygen needed to oxidise the lactic acid that has built up in muscles during anaerobic respiration.

During a sprint, athletes expend so much energy so quickly that they must rely on anaerobic respiration. Their muscles have to work so hard that the blood simply cannot supply their muscles with oxygen quickly enough. This is possible for the duration of the race, but at the end of the race their muscles will contain a lot of lactic acid. This must be broken down otherwise it will start to damage their muscles. Lactic acid is oxidised by oxygen to carbon dioxide and water. The carbon dioxide is then excreted as if it had been produced by aerobic respiration. The amount of oxygen needed to oxidise the lactic acid is called an oxygen debt. The sprinters in Figure 7.16 are paying back their oxygen debt.

Figure 7.16 These sprinters have produced a lot of lactic acid in their muscles during the race. What happens to the lactic acid?

Test yourself

33 a 'Anaerobic respiration is far less efficient than aerobic respiration.' Explain what this statement means.

b How does your body respond when you cycle up a steep hill quickly?

c Once you finish cycling up the hill, you need a rest. Explain what happens in your muscles while you are resting. Use all the key words in the definition boxes in your answer.

Chapter 8
How do our bodies pass on characteristics?

ICT

In this chapter you can learn to:

- use the internet to find animations of meiosis and mitosis
- have a debate with a wide range of students using the school VLE

Setting the scene

All the information about our features and characteristics is inherited from our parents, and we start as a single fertilised cell. How can you develop from just one cell?

Practical work

In this chapter you can learn to research easy-to-recognise genetically inherited features of humans and carry out a survey.

B2 8.1

Learning outcomes

- Understand that Mendel planned his experiments.
- Understand that Mendel carried out and recorded carefully a large number of experiments
- Understand that this produced sufficient data to analyse to give reproducible results.

Figure 8.1 Gregor Mendel at work.

Test yourself

1. What does the word 'inheritance' mean when it is used in science?
2. What prompted Gregor Mendel to carry out his research into inheritance?
3. What did Mendel do to ensure that his findings were reproducible?
4. Why was it important for Mendel to obtain reproducible results?
5. Explain how Mendel's results showed that the blending theory of inheritance was unsatisfactory.

How did an Austrian monk help to shape our understanding of inheritance?

We know today how the features and characteristics that we inherited from our parents are passed on, but this was not always the case. It was only through careful experimentation that scientists gained this understanding. In this section, you will learn about the key developments that led to our current understanding of inheritance.

For many years people believed that the characteristics of parents combined in some way when they had children. For example, if a mother had black hair and a father had blonde hair, the two colours would combine to produce brown-haired children. This was also thought to be true of animals and plants. It was known as the 'blending theory' of inheritance.

An Austrian monk called Gregor Mendel, who was born in 1822, disagreed with this theory. Mendel worked in the gardens of an Austrian monastery and noticed that pea plants had either purple or white flowers. He sowed the seeds produced from these flowers. The new plants that grew only ever had purple or white flowers (the same colour as the two parents), never light mauve. He concluded that the parent plants passed on specific features to the offspring plants and not combined features.

Meticulous and mathematical are two words that can be applied to Mendel's approach to the problem. Although Mendel had not trained as a scientist, he realised that he needed to carry out controlled experiments. He could then collect reproducible evidence to analyse mathematically to prove his theory. He started by doing some preliminary experiments in which he pollinated purple and white flowered plants together. From this he confirmed that when these two types of pea plants were crossed they produced plants with white flowers and plants with purple flowers but never any plants with colours in between.

For his main experiment, Mendel took a purple flowering plant that had produced some white flowered plants, and a white flowering plant. He then pollinated these two plants together for several generations and counted the numbers of purple and white flowering plants that were produced.

Mendel analysed his results and discovered that for every three purple flowering plants there was only one white flowering plant, a ratio of 3:1. From his results, Mendel concluded that some inherited characteristics, such as purple flowers, had a stronger influence over the offspring. He called these 'stronger' influences dominant and the 'weaker' influences recessive. These findings showed clearly that the blending theory was insufficient to explain inheritance.

Mendel published his findings in 1866, expecting people to appreciate that inheritance could be much better explained using his ideas. Sadly, his ideas were largely ignored for 34 years until three other researchers published similar findings based on their own experiments. At the same time, Mendel's work was translated into English and people started to take his ideas seriously. They realised that Mendel had provided a far better explanation of inheritance than the blending theory. With the development of more powerful microscopes that enabled the chromosomes to be seen in the stages of cell division, we realise what great insight Mendel had. Mendel's theories fit and are used in the construction of genetic diagrams.

Stem cell research in the news

HOW SCIENCE WORKS ACTIVITY

Stem cell treatment from cord-blood for blood disorders — acute myeloid leukaemia (2010)

Stem cell experiment lets diabetics forgo insulin

Risky transplants performed on 13 young diabetics in Brazil (2007)

Doctors have launched a trial to test whether heart disease can be treated using a patient's own stem cells (2005)

Stem-cell transplants from own bone marrow may control and even reverse multiple sclerosis symptoms if done early enough, a small study has suggested (2009)

Figure 8.21

These are just a few recent headlines from articles about experimental research that focuses on human stem cells. Of all the current biomedical research, stem cell research has probably attracted more attention than any other from the media. But why is this?

To understand the level of interest and range of views expressed, it is crucial to know exactly what happens when stem cells divide and the potential medical benefits these cells could bring.

1. Explain why stem cells are essential in human development.
2. Nerve cells, red blood cells, muscle cells and pancreatic cells are some of the specialised cells that stem cells produce. Each one has a specific structure that allows it to carry out its function. Find out and state the function of each of these specialised cells. Explain how each cell's structure enables it to carry out its function.
3. Stem cells can be collected and kept alive outside the body. With this in mind, suggest two possible uses scientists could make of stem cells. For each suggestion explain how you think scientists would carry out the procedure. Put the main points into a flow diagram.
4. In small groups, debate reasons that people may have to support or oppose embryonic stem cell research. Write these down as a list of reasons for, and a list of reasons against the research.
5. Now put the reasons in each list into order based upon which ones present the strongest arguments for and against stem cell research.
6. Finally, write a paragraph expressing your views about this issue. Support your views with reasons.

B2 8.7 How organisms change through time

Learning outcomes

- Know that new species can be formed by geographical isolation.
- Understand how fossils are evidence for evolution.
- Understand the different ways in which fossils may be formed.

Have you ever found a fossil and wondered just how old it was? This leads to thoughts of how life began on Earth. Scientists have many hypotheses. Overall, though, there is little valid and reproducible evidence. It is thought that the first life forms were bacteria and unicells, followed by soft-bodied animals. Fossils of soft-bodied animals are rarely found because they are usually eaten or decay quickly. But in May 2010, 1500 soft-bodied marine animal fossils of several sorts were found in Morocco. These well-preserved fossils date back to nearly 500 million years ago and fill in a lot of gaps in the record.

How were fossils formed?

When an organism dies, it is normally eaten by other animals or decomposed by bacteria and fungi quickly, leaving nothing to see except possibly a few bones.

Normal fossil finds are from the hard parts of organisms such as bones and shells. Fossils were formed when an organism fell into a marsh or bog in which the conditions were not suitable for decay — that is, no oxygen and water with a low pH. The remains were buried under layers of vegetation and became compressed with age. (Many human remains from the Iron Age have been found in peat bogs. Although not fossils, they show how effective these conditions are for preservation.) Alternatively, fossils can also form when organisms have been covered with layers of sand, volcanic ash or silt. They become compressed and impregnated with mineral salts from water, turning them into stone. There are other interesting remains such as insects preserved in tree resin called amber, dinosaurs' bones, footprints and eggs, petrified trees, remains in ice such as mammoths and burrows in mud deposits.

Test yourself

24 Explain three ways in which fossils were formed.

25 You are examining the layers in a chalk cliff. Where will the oldest fossils be found?

Figure 8.22 A peat bog body. No oxygen and weakly acidic conditions prevent decay.

Figure 8.23 The Grand Canyon, Arizona. A 500 million year time sequence in the rock layers.

What evidence do scientists get from the fossil record that can be related to evolutionary theory?

Scientists study fossils to find out about the animals and plants of the past. Fossils can show the time sequence of existence. In sedimentary rocks the material is laid down in layers so the deeper you go down, the older the material and the older the fossils. The Grand Canyon in Arizona has a depth of 1700 m and this represents a time span of 500 million years. Geologists can trace changes in climate and rock movements and fossil experts can follow changes in shape and structure of plants and animals that could result from climate changes.

Evidence from fossils

There is a definite order in which fossils appear.

1 The bottom layers (oldest rocks) show only a small variety of marine invertebrates and represent several million years without much change.

Chapter 9 Why are diffusion and active transport so important to living things?

Setting the scene

For cells to be able to function properly there must be correct uptake of nutrients and water, and efficient removal of waste products. These processes can be applied in medicine, horticulture and sports science and are investigated in this chapter.

ICT

In this chapter you can learn to identify which types of multimedia make it easier to inform people about isotonic drinks.

Practical work

In this chapter you can learn to:

- identify anomalous results and understand why they might occur
- draw conclusions from the results of an experiment on osmosis
- design an experiment to investigate the effect of humidity on the rate of transpiration

B3 9.1 Osmosis

Learning outcomes

- Know how water moves in and out of cells by osmosis.
- Understand how osmosis is affected by the concentrations of the solutions inside and outside a cell.

Having the correct amount of water in your blood is crucial, and this is controlled by your kidneys. If the concentration of water in your blood changes, it will affect the cells in your body because water can diffuse into and out of cells. The diffusion of water into or out of cells is called **osmosis**. This is illustrated very clearly by red blood cells.

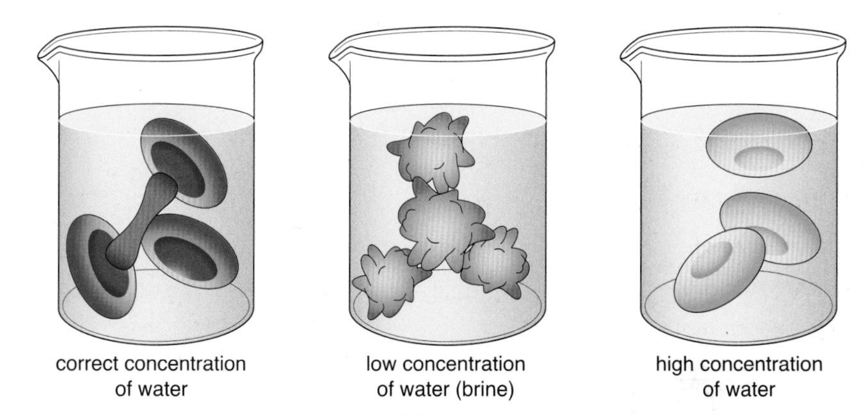

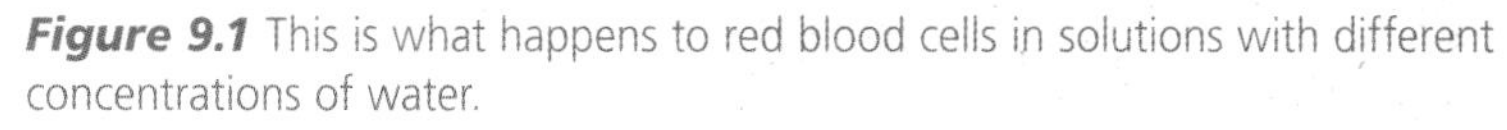

Figure 9.1 This is what happens to red blood cells in solutions with different concentrations of water.

Osmosis is the diffusion of water, through a partially permeable membrane, from a region of higher water concentration to a region of lower water concentration.

Red blood cells are specialised cells with a large surface area and no nucleus — a structure that makes them very good at carrying oxygen around the body. When they are in a solution with the correct concentration of water, they have a recognisable 'biconcave' shape. If the concentration of water around them is too low, they shrivel up; and if it is too high they swell up and can even burst. You can see a similar effect if you put a tomato in a glass of water and another in a glass of water with a lot of salt dissolved in it.

Test yourself

1 Explain, in terms of water movement, what has happened to the red blood cells in each solution in Figure 9.1.

2 Visking tubing is a synthetic material with tiny holes in it. These holes are big enough for small molecules, such as water, to pass through.
  a If a length of Visking tubing is filled with pure water and then placed in a beaker of sugar solution, what will happen?
  b Explain your answer to part a.

What can affect osmosis?

To fully understand osmosis, you have to consider the water molecules and any substances dissolved in the water, both inside and outside the cell. Look at Figure 9.2 and think about the number of water (blue) and glucose molecules (red) in each solution.

In Figure 9.2 (a), water molecules move from left to right, from a region of higher water concentration to a region of lower water concentration. Note that some people explain this by referring to the concentration of glucose in

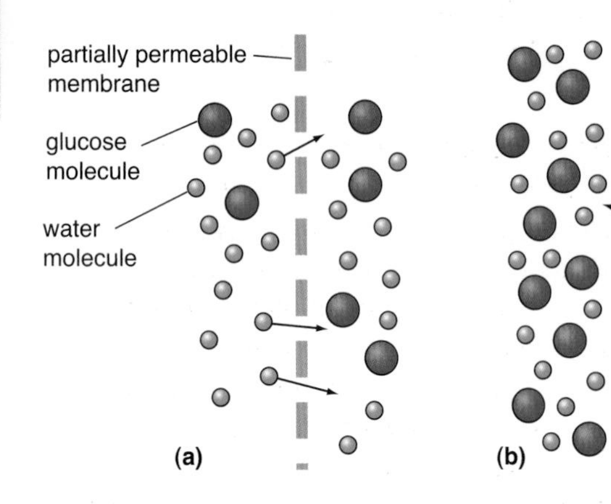

each solution. In this case, you would say that water moves from a less concentrated (dilute) glucose solution to a more concentrated glucose solution through a partially permeable membrane. In Figure 9.2 (b) the water molecules move from right to left, because the concentration of glucose is higher on the left than on the right. So, you can change the movement of water by osmosis by adding more water or by adding more solute to the solution.

Figure 9.2 Two partially permeable membranes. Both membranes have the same concentration of solution on the right-hand side. However, water moves in opposite directions through the membranes.

Test yourself

3 For each diagram in Figure 9.3, state which way the water will move, and explain why this will happen in terms of water and solute concentration.

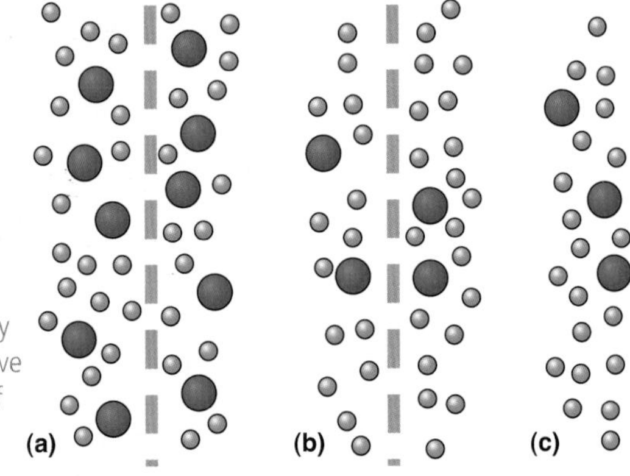

Figure 9.3 These partially permeable membranes have different concentrations of solution on each side.

Changing the rate of osmosis

HOW SCIENCE WORKS
PRACTICAL SKILLS

Two students carried out an investigation using cylinders of potato. They cut out five potato cylinders and recorded the mass of each one. They made up five glucose solutions with different concentrations, and left a potato cylinder in each solution for 24 hours. After 24 hours they dried each potato cylinder and measured its mass again. The results are shown in Table 9.1.

Glucose concentration in moles per litre	Starting mass of potato cylinder in grams	Mass of potato after 24 hours in grams
0	7.76	9.38
0.2	8.01	8.04
0.4	7.96	7.21
0.6	7.92	7.78
0.8	7.93	6.65

Table 9.1 The results from the five potato cylinders left in different glucose solutions for 24 hours.

1. Why did the students dry each potato cylinder before measuring its mass after each experiment?
2. What variables should be kept the same to make this investigation a fair test?
3. Work out the change in mass of each potato cylinder. These mass changes will be positive for cylinders that got heavier and negative for those that got lighter.
4. Now plot a graph for these data showing the glucose concentration on the *x*-axis and the change in mass on the *y*-axis. Think carefully about how you set this out so that you can plot positive and negative values on the same graph.
5. One of the results appears to be anomalous. Which one is it and what may have caused this?
6. Write a conclusion for this investigation. You will need to describe the pattern shown by the graph and use scientific ideas to explain what happened.
7. Use the graph to predict the concentration of glucose that you would use if you wanted no change in mass of a potato cylinder. Explain, in terms of osmosis, why this would happen.

B3 9.2

Learning outcomes

- Be able to explain how dissolved substances move by diffusion and active transport.
- Understand that active transport requires energy, whereas diffusion does not.

What is the difference between diffusion and active transport?

You should know how gases and dissolved substances move by diffusion. The particles in a gas or a liquid are constantly moving. If there is a high concentration of particles in one part of a gas or solution, they will tend to spread out and move to where there is a lower concentration of these particles. This movement of particles from a region of higher concentration to one of a lower concentration is called **diffusion**. In humans, oxygen diffuses into cells from the blood through cell membranes so that the cells can respire. Oxygen is at a relatively high concentration in the blood. It moves by diffusion to where it is at a lower concentration inside the cell. In moving from a place of higher concentration to one of lower concentration, we say the oxygen is moving down a **concentration gradient**.

Diffusion is the movement of particles in a gas or solution down a concentration gradient.

A **concentration gradient** exists between two areas when one area has a higher concentration of a substance than the other area.

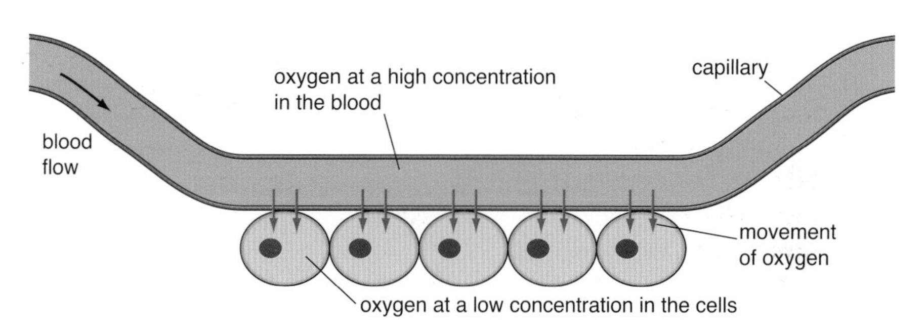

Figure 9.4 Oxygen molecules move down a concentration gradient, from a higher concentration in the blood to a lower concentration in any respiring cell. Can you give some examples?

Active transport

When diffusion occurs in living things, substances move from regions of higher concentration to regions of lower concentration — this does not require cellular energy. However, living organisms sometimes need to move a substance from an area where its concentration is low to an area where there is already a higher concentration of the substance. Think about a plant taking minerals into its roots from the soil. Some minerals, such as phosphate ions, have a very low concentration in soil and a higher concentration inside the root cells. However, the plant still needs to take in phosphate ions from the soil. In this situation, diffusion is useless because the phosphate ions would move out of the root cells into the soil, down a concentration gradient. Instead, the plant uses a process called **active transport** to move phosphate ions from the soil into the root cell.

Active transport is the movement of particles through a membrane, against a concentration gradient. The energy required for active transport is obtained from cellular respiration.

Active transport is used in living organisms to move substances through a membrane against the concentration gradient, from a lower to a higher concentration. Unlike diffusion, active transport cannot rely on the random movement of particles to move them through a membrane. Instead, chemicals in the membrane bond with particles on the outside of the cell, move them across the membrane and then release them into the cytoplasm. This process uses energy that the cell releases during respiration. Humans use active transport to absorb sugar from the small intestine, where there is a very low concentration of sugar, into the blood, where there is a higher concentration.

Test yourself

4 Why do substances *not* move through visking tubing by active transport?

5 When plants photosynthesise they produce sugar inside cells, and from here the sugar is transported around the plant. The sugar is often at a lower concentration inside the cells than in the tubes that transport it around the plant. However, plants can move the sugar from their cells into the transport tubes.
  a Describe the concentration gradient that exists between the plant cells and transport tubes.
  b How do you think the sugar is moved from the cells into the tubes so that it can be transported around the plant?

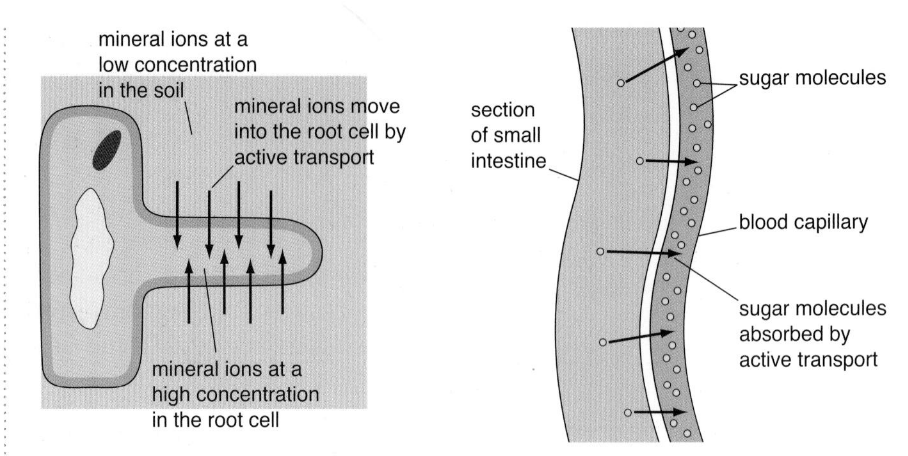

Figure 9.5 Can you explain in detail what is being shown in these two diagrams?

One way to understand the difference between diffusion and active transport is to compare them to riding a bike downhill and uphill. If you are on a bike at the top of a hill, you can roll down without using any of your own energy — this is like diffusion; you use no energy and move down the gradient (slope). In contrast, active transport is like riding a bike uphill — you are cycling up a gradient and have to use some of your own energy. You obtain this energy from respiration, just like a cell using active transport.

Humans, water and mineral ions

Water balance in humans is very important and we have a tendency not to drink enough fluids. 75% of the body mass is water — the water in the body is in cells, blood plasma and tissue fluid, which surround all cells. If you are dehydrated, the concentration of the blood plasma and tissue fluid is high, and this causes water to pass out of cells by osmosis. This means that the cells cannot function efficiently. The body monitors your blood concentration and if you begin to dehydrate you feel thirsty or dry-mouthed, light-headed and less alert. If you are thirsty the best drink is water but you may reach for some fruit squash, which includes some sugar (unless it is sugar-free).

If you are exercising seriously and training hard enough to sweat for a long period of time (over 90 minutes) you are losing water and mineral ions — these need to be replaced. An isotonic sports drink contains a 6–8% carbohydrate solution, 50 mg of sodium per 100 cm^3, along with smaller amounts of the other ions, such as potassium and chloride, which are lost in sweat. The word 'isotonic' means the ions are at the same concentration as body fluid. This drink is absorbed quickly to replace lost water and ions, and the sugar provides a little energy.

Figure 9.6 An athlete using an isotonic sports drink. What is being replaced by this drink?

Test yourself

6 Why is it important to avoid getting dehydrated when you are exercising?

7 Marathon runners can occasionally drink too much water and become overhydrated. This is more likely to happen when they have been sweating heavily and have lost a lot of salt.
  a What effect will overhydration have on the cells in the body?
  b Why does losing a lot of salt from the body increase the effect of overhydration on body cells?

Test yourself

8 Sports drinks can be divided into three types. *Hypotonic drinks* have a sugar concentration lower than your body fluid; *isotonic drinks* have the same concentration of sugar as your body fluid; *hypertonic drinks* have a higher sugar concentration than your body fluid.

a Which type of sports drink would be best to drink if you wanted to rehydrate your cells quickly?

b Explain your answer to part (a).

c What are the advantages and disadvantages of drinking hypertonic drinks if you go for a long bike ride on a hot day?

Isotonic sports drinks should not be confused with 'energy drinks', which contain high levels of sugar and caffeine and are not for training use. If you look at labels, you might be surprised to find how much sugar is in some drinks — and that certainly explains weight gain and tooth decay. Coffee, tea and alcohol tend to make the kidneys remove water from the body.

Isotonic sports drink — ICT

Create a resource to explain the difference between an isotonic drink and an energy drink to young people.

Create a multimedia advertising campaign for an isotonic sports drink, explaining why it is effective.

B3 9.3 How are our breathing and digestive systems specialised for absorbing substances?

Learning outcomes

- Be able to describe how the lungs are specialised to absorb oxygen and dispose of carbon dioxide.
- Know that the lungs are protected by the ribcage in the thorax, and separated from the abdomen by the diaphragm.
- Understand lung ventilation.
- Know that the villi are specialised for the absorption of soluble end-products of digestion.

Your breathing and digestive systems both have the ability to transfer substances through membranes. The breathing system takes air into the lungs so that oxygen can be absorbed into the bloodstream. At the same time, carbon dioxide passes out of the bloodstream into the lungs before being exhaled. The digestive system absorbs nutrients from your food into the bloodstream. Both systems have specialised features that increase their ability to transfer materials through their surface membranes.

The breathing system

Your breathing system moves gases into and out of your body — see Figure 9.7. When you breathe in, your lungs inflate with air. When you breathe out, most of

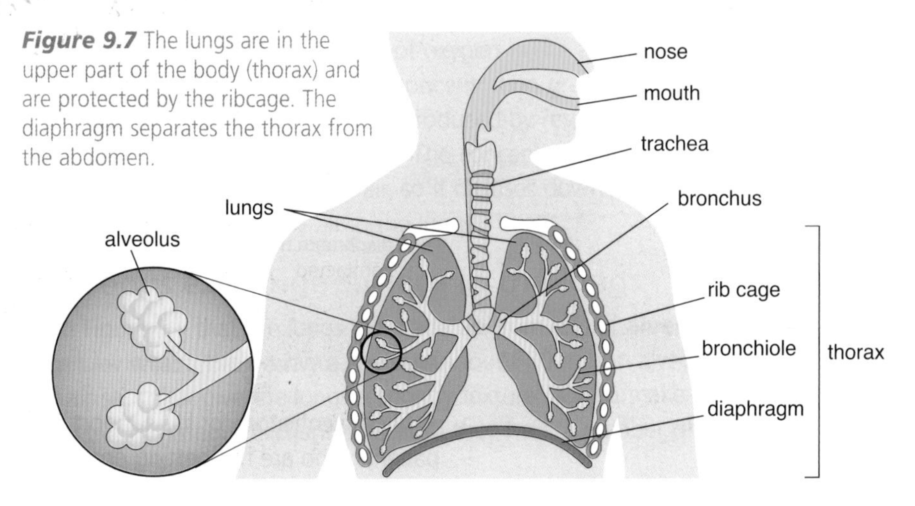

Figure 9.7 The lungs are in the upper part of the body (thorax) and are protected by the ribcage. The diaphragm separates the thorax from the abdomen.

What affects the rate of transpiration in plants?

HOW SCIENCE WORKS
PRACTICAL SKILLS

The rate of transpiration in plants can be measured using a potometer (Figure 9.20). The plant draws up the water it needs through a very narrow tube. An air bubble in the tube moves as the water is absorbed. By measuring how far the air bubble moves each minute you can compare the transpiration rates under different conditions.

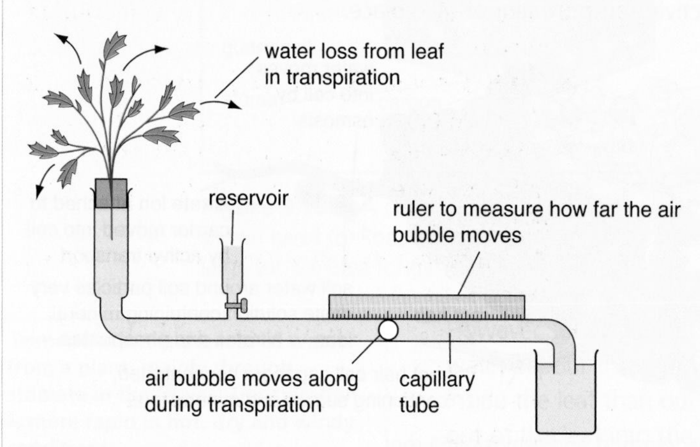

Figure 9.20 Measuring the rate of transpiration with a potometer.

Zoe carried out an investigation using a potometer to find out how different conditions affected the rate of transpiration. She carried out three experiments under the following conditions:

Ⓐ on a bench in the laboratory

Ⓑ in the laboratory with a fan blowing over the plant

Ⓒ in a warm incubator.

Her results are shown in Table 9.2.

	Distance the bubble moved after each minute in mm									
Time in min	1	2	3	4	5	6	7	8	9	10
Experiment A	2	3	5	7	10	11	13	15	18	20
Experiment B	13	24	38	49	73	75	89	104	118	132
Experiment C	5	12	18	24	30	35	41	47	53	58

Table 9.2

1. What should Zoe do to check that her results are accurate?
2. Plot a graph of Zoe's results. Choose the scales so that all the results can be shown on one graph.
3. What conclusions can you draw about the effect of different conditions on the rate of transpiration?
4. Sketch on your graph the line you would expect to get if you carried out the experiment using a hair drier set to 'warm' instead of just a fan in Experiment B.
5. One of the results appears to be anomalous (doesn't fit with the others). Which one is it? Suggest one reason for the anomalous result.
6. Design an experiment to investigate the effect of humidity on the rate of transpiration. ('Humidity' means the amount of water vapour in the air.)
7. What effect do you think increasing the humidity will have on the rate of transpiration?

Homework questions

1 Look at Figure 9.16 (b), showing water and mineral ions being absorbed into a root hair cell. Water is moving from an area of high water concentration in the soil to a lower one inside the root hair cell. On the other hand, minerals must often move from an area of lower concentration in the soil to one of higher concentration inside the cell.

- **a** How do root hair cells increase the rate at which a plant can absorb water and mineral ions?
- **b** What process is used to absorb water into root hair cells?
- **c** How are mineral ions absorbed when they are at a lower concentration in the soil than in the root hair cell?

2 Leaves lose water by transpiration.

- **a** Why does this occur slowly on a cloudy day?
- **b** Why is transpiration faster on a bright, windy day?

3 A leafy shoot was set up as shown in Figure 9.20. The time taken for the air bubble to move along the capillary tube was recorded. The diameter of the capillary tube was 2.0 mm.

Table 9.3 below shows the results obtained under different conditions.

- **a** Copy and complete these two columns in the table — distance moved and distance moved per minute.
- **b** The final column is to show the volume of water absorbed by the shoot. Calculate each of the volumes — use the formula $\pi r^2 h$, where h is distance moved along the tube and r is the radius of the tube.
- **c** Which factors involved in this experiment cannot be quantified?
- **d** Which pairs of results give valid comparisons? Discuss these.
- **e** What general conclusion can you make from these results?
- **f** A student thought that he would get a really fast result by keeping the temperature high and using a fan. The bubble stopped moving after a few millimetres. Can you explain this?

Table 9.3

Air movement	Temperature in °C	Humidity	Start position in mm	Final position in mm	Distance moved (h) in mm	Time taken in min	Distance moved in mm per minute	Volume of water taken up
Still	20	Dry	4	54	50	112		
Slight breeze	20	Dry	3	55		75		
Still	25	Dry	7	60		55		
Still	20	Damp	2	50		150		
Slight breeze	20	Damp	3	55		80		
Still	25	Damp	4	56		70		

Exam corner

A student set up the following experiment to compare water loss by plants at different light intensities. Each of three pot plants was covered by a clear plastic bag tied securely around the stem. Plant X was placed in bright sunlight, plant Y was enclosed in a black plastic bag as well, plant Z was placed in low light. Inside each bag was a piece of dried cobalt chloride paper, which was pale blue. The time taken for the cobalt chloride paper to turn pink was recorded. (Cobalt chloride detects water vapour and turns from pale blue when dry to pink.)

The cobalt chloride paper in bag X changed first, in bag Z second — that in bag Y did not change.

a Suggest a reason for the order of change. *(2 marks)*

b Explain one reason why you think this was or was not a reproducible experiment. *(2 marks)*

Student A

a Plant X was photosynthesising in the bright light and so the stomata would be wide open and water vapour would have escaped. ✓ Plant Z would have its stomata half open because there was not so much light. Plant Y could not photosynthesise in the dark so the stomata were shut, so no water vapour could escape. ✓

b Yes, it gave the results expected. ✗ ✗

Examiner comment Student A has spotted the link between stomatal opening and transpiration in part (a) and has given a good answer. In part (b), when setting up an experiment to try to prove something, you should consider all the factors. What variables would you control — leaf surface area, temperature, were the plants hydrated before the experiment?

Student B

a Plant X was transpiring fastest and plant Y was slowest. Transpiration is controlled by light and Y had no light. ✗ ✗

b No. ✓ Temperature was not controlled so plant X in the bright sunshine would also get hot. ✓

Examiner comment Student B just repeats the order given in the question in part (a) — copying down what a question says does not gain marks — and relates transpiration to light without any explanation. So, no marks — mention of stomata opening is needed. In part (b), temperature has been spotted as an uncontrolled variable, suggesting the experiment was not reproducible.

Chapter 10
How materials are transported around an organism

Setting the scene

Knowledge of the functioning of the heart has led to the development of several methods for helping to maintain active life. What do you think this image shows?

ICT

In this chapter you can learn to evaluate animations to find some that help you learn about the heart.

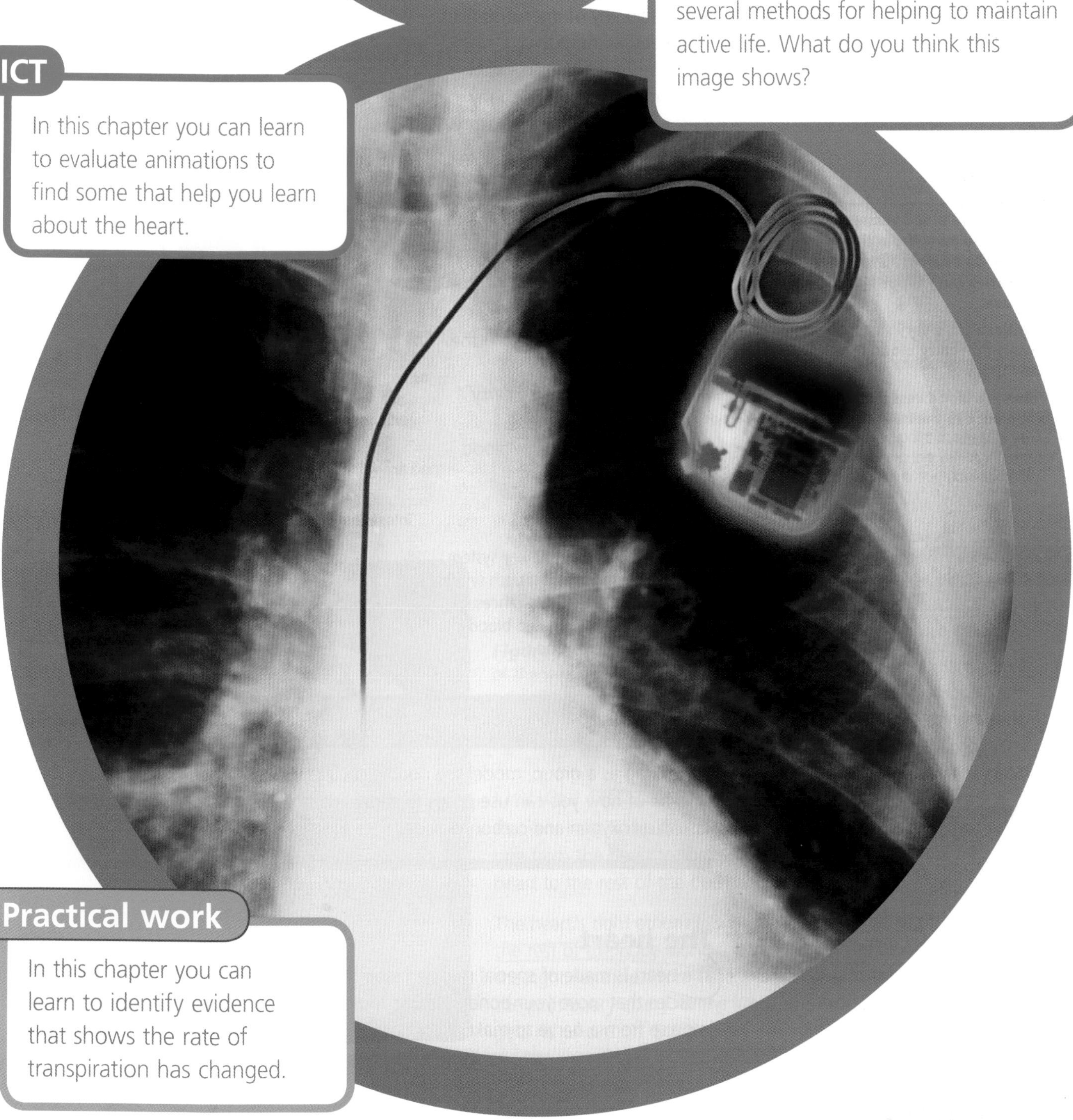

Practical work

In this chapter you can learn to identify evidence that shows the rate of transpiration has changed.

Red blood cells

It is crucial that all your cells receive a good supply of oxygen. Oxygen can dissolve in plasma, but this way of transporting oxygen would not supply sufficient oxygen to keep all your cells functioning. However, red blood cells have special features that make them incredibly efficient at transporting oxygen:

- they have a large surface area due to their biconcave shape
- they are packed with a substance called haemoglobin
- they have no nucleus, so there is a greater volume available for **haemoglobin**, and hence for carrying oxygen
- the cell membrane is thin so oxygen can diffuse in easily

Haemoglobin is the chemical in red blood cells that combines with oxygen — to form oxyhaemoglobin. Oxyhaemoglobin releases the oxygen to cells in the body.

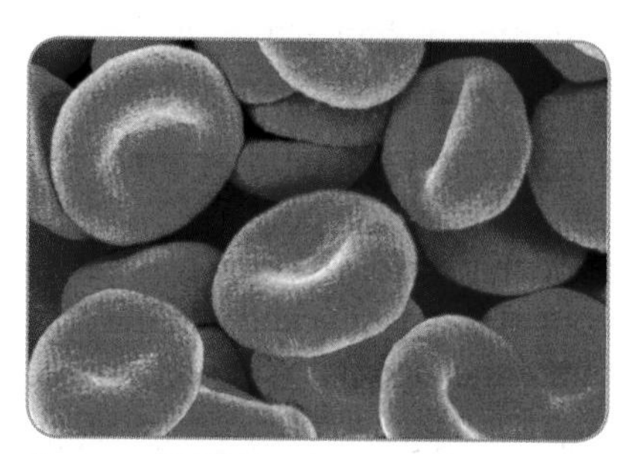

Figure 10.14 List the ways that red blood cells are adapted for carrying oxygen.

Haemoglobin is a substance that can combine with oxygen reversibly. When blood passes through the lungs, where oxygen is in high concentration, oxygen diffuses from the alveoli into the blood (Figure 10.15) where it combines with haemoglobin molecules to form oxyhaemoglobin. When the oxyhaemoglobin reaches other organs that have a lower concentration of oxygen, the oxygen splits from the oxyhaemoglobin and diffuses into the organ's cells, where it is used for respiration. This leaves haemoglobin in the blood that can combine with more oxygen when it next passes through the lungs. The lungs are the only place where haemoglobin can load up with oxygen for the purpose of delivering it to respiring cells.

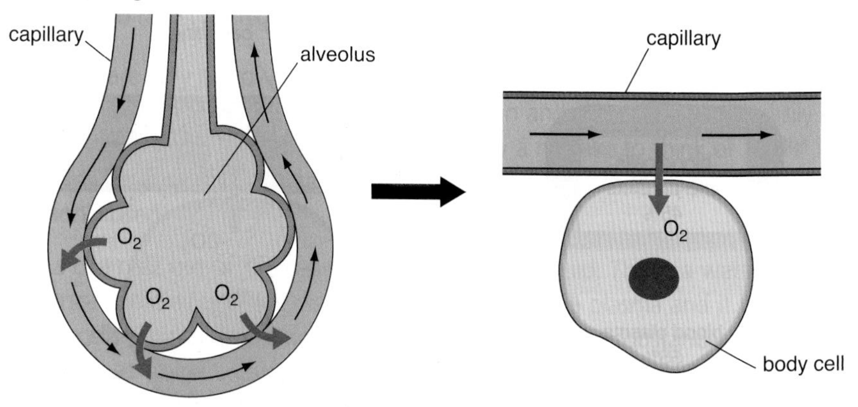

Figure 10.15 Haemoglobin transports oxygen from the lungs to other organs as oxyhaemoglobin in the blood.

Test yourself

10 How are red blood cells adapted to absorb and carry oxygen?

11 Our bodies use iron in our food to make haemoglobin. People who have a deficiency of iron in their diet often complain of tiredness. Explain why you think this happens.

12 Why is it dangerous to travel in a car with a leaky exhaust pipe?

Some people suffer from carbon monoxide poisoning as a result of faulty gas fires and boilers. Carbon monoxide is a gas that can combine with haemoglobin just like oxygen. In fact, carbon monoxide combines about 240 times more strongly with haemoglobin than oxygen does. It forms carboxyhaemoglobin, which is stable and prevents oxygen combining with the same molecules of haemoglobin. This greatly reduces the blood's ability to carry oxygen. As a result, the brain and body cells receive insufficient amounts of oxygen, and this can lead to unconsciousness and even death. This is why it is so important that gas fires and boilers are serviced regularly and that carbon monoxide detectors are fitted.

White blood cells

White blood cells are an important part of the body's defence against disease. There are two main types of white cells — both are involved in attacking and destroying pathogens. Unlike red blood cells, white blood cells have a nucleus. White cells that have a large, round nucleus react to a pathogen by producing protein antibodies that are specific to the markers on the pathogen surface, and cause the pathogens to stick together. White blood cells that have a lobed nucleus ingest pathogens that have been stuck together by antibodies, taking them into a vacuole. The pathogens are chemically digested by enzymes produced in the white blood cell cytoplasm, which are added into the vacuole.

Platelets

Platelets are very small structures with no nucleus or DNA but they do have some enzymes in their cytoplasm. Their important function is in blood clotting at the site of a wound.

Artificial blood products

Blood is an important substance used in surgery, after accidents and for many types of treatment. In the UK, it is currently collected from voluntary blood donors. The main problem is that it has a short shelf-life so there is always a need for more donations. Currently about 75 million units of blood are used worldwide in a year. For a long time, experiments have been carried out to try to make artificial blood — in particular, the widespread problem of transmission of virus infections (e.g. HIV) in blood has increased the need for large quantities of acceptable blood. The requirements of a blood substitute are that it must not cause rejection, it must have a long shelf-life and be able to be kept and transported, it should not transmit infections, such as HIV, and it should be good at transporting oxygen.

So far, no successful substitute for haemoglobin has been manufactured. A substance called PolyHeme came close. PolyHeme got to phase III clinical trials but there were found to be more heart attacks among accident victims treated with this product than with the existing treatment. The product was withdrawn in 2010. Artificial blood will certainly be a good selling product. In the UK there are currently experiments to make artificial haemoglobin and also to make artificial blood from stem cells.

Test yourself

13 Suggest some reasons why artificial blood would be a useful product.

Blood substitute — ACTIVITY

If you were considering making artificial blood, what functions would it need to fulfil and what properties should it have?

Look online to see how your suggestions match those being tried.

Chapter 11
How our bodies keep internal conditions constant

Setting the scene

The hypothalamus region of the brain (H) monitors many internal factors, including temperature and water content of the blood. Medical knowledge has led to the development of artificial control mechanisms. When vital organs such as the kidneys or the pancreas fail, sufferers can still lead normal lives.

H

Practical work

In this chapter you can learn to make careful observations and decide which variables are the most important in an investigation on cooling.

ICT

In this chapter you can learn to:

- select the most effective way to present information on how the body gets rid of waste
- create an online debate on ethical issues surrounding kidney transplants
- research up-to-date information on diabetes

B3 11.1

Learning outcomes

- Understand how the body gets rid of waste products.
- Know how carbon dioxide and urea are removed from the body.
- Understand that excess or worn out proteins are broken down to form urea.

Homeostasis is the maintenance of steady conditions within the body, including water content, blood glucose concentration and body temperature.

How does your body control its internal conditions?

Your body is constantly monitoring its internal conditions and making changes to try to control them. This means that your body is controlling the conditions for your cells, blood and body tissue. The internal conditions that are controlled include the water and ion content of your body. Your temperature, oxygen, carbon dioxide and blood sugar levels are also controlled. **Homeostasis** is the name given to mechanisms in the body that regulate its internal conditions, keeping them within narrow ranges.

How does your body get rid of waste products?

Some of the chemical reactions that take place in your body produce useful products such as proteins. Respiration is essential because it releases energy from glucose. However, these reactions also produce waste products that can be toxic to the body. Fortunately your body has ways of removing these waste products so that they do not reach harmful levels in the blood.

How is carbon dioxide removed from the body?

Respiration is a series of chemical reactions that take place in cells. It is summarised by the equation:

glucose + oxygen → carbon dioxide + water (+ energy)

The carbon dioxide that is produced moves from the cells into the blood by diffusion. Too much carbon dioxide in the blood can be dangerous because it lowers the blood pH (makes it more acidic) and inhibits enzyme action. As blood flows through the lungs, carbon dioxide diffuses out of the blood and into the air in the lungs, to be breathed out in the next exhalation. At the same time, oxygen diffuses from the air into the blood. This takes place across the walls of the alveoli in a process called gaseous exchange.

Test yourself

1 Explain why oxygen diffuses from the alveoli in the lungs to the blood, and why carbon dioxide diffuses from the blood into the alveoli.

2 Why does your breathing rate increase when you go for a run? Refer to both oxygen and carbon dioxide in your answer.

How are other types of waste removed from the body?

Enzymes called proteases digest proteins, breaking them down into amino acids. The body uses amino acids to make all the proteins that you need for growth and to control body processes. You often have more amino acid molecules in your blood than your body can use. The liver converts these excess amino acids into a chemical called urea. Urea can harm the body if its concentration is too high, so it must be removed. Figure 11.1 shows how this happens.

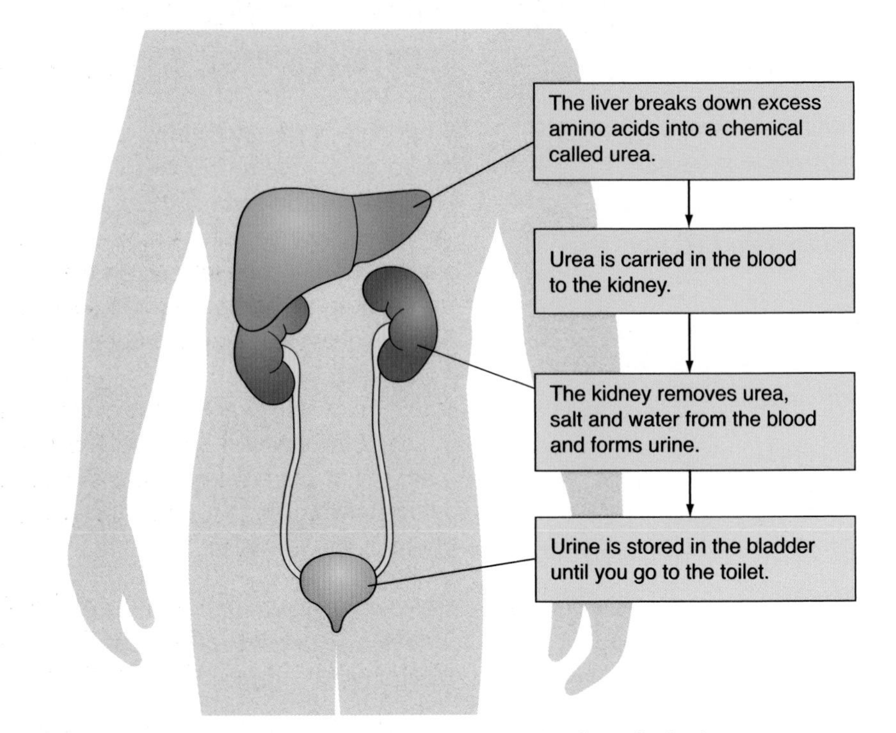

Figure 11.1 The removal of excess or worn out proteins from the body.

As you can see in Figure 11.1, the liver and the kidney work together to remove amino acids from the body. The liver breaks down excess amino acids in a series of chemical reactions. This produces urea, which dissolves in the blood plasma and is then transported to the kidneys. Because it is dissolved in the plasma, it is removed from the blood along with some of the water it is dissolved in. This mixture, along with salt, makes up the urine that is stored in the bladder until you go to the toilet.

Test yourself

3 What is the difference between urea and urine?

4 People who suffer from liver damage are sometimes asked to reduce the amount of protein in their diet. Explain the scientific reason behind this advice.

5 Urine can be tested for the presence of amino acids. What possible medical problems do you think somebody might have if amino acids are detected in their urine?

Waste disposal

ICT

Create a presentation, animation or podcast on what waste products the body produces and how it gets rid of them.

B3 11.2

Learning outcomes

- Be able to describe how the kidneys remove toxic substances from the blood.
- Be able to explain how the kidneys maintain the correct concentrations of water, sugar and dissolved ions in the blood.

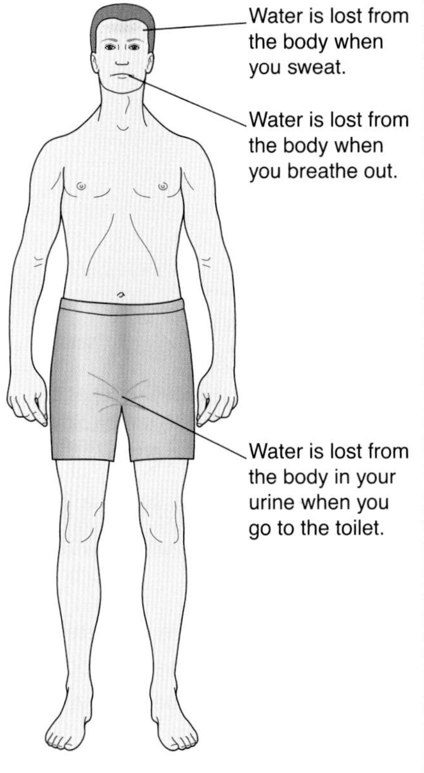

Figure 11.2 Ways in which water is lost from the body. Will the volume of water lost be constant? What might cause it to change?

How do the kidneys help to maintain internal conditions in the body?

Controlling the water and ion content of the body

When you get too hot, you sweat to reduce your body temperature. The water in the sweat evaporates from your skin, cooling you down. But sweat doesn't just contain water. Ions, such as sodium and potassium, which are dissolved in your sweat, are also lost. Both water and ions are essential for good health and we can obtain both when we eat and drink. Remember the effects of dehydration and over-hydration. Too much water in the blood can cause body cells to swell up, or even burst. Too little water causes cells to shrivel. Either way, too much or too little water in our cells means that they stop functioning properly — body tissues, such as muscle, can stop working and cells can be permanently damaged.

In severe cases, dehydration or over-hydration can lead to unconsciousness or even death. Altering the concentrations of the ions dissolved in the blood also interferes with water uptake by the cells.

How does a hormone control the water content of the body?

It is essential to have the right amount of water in your body to keep the concentration of the blood constant. But the volume of water consumed when you eat and drink varies. So too does the water lost from your body when you breathe out, sweat and urinate. On a hot day, you will lose more water by sweating. If you don't drink enough you will become dehydrated and your body will make changes to save the water. Figure 11.3 shows how blood water content is controlled by one hormone.

Blood concentration is monitored in the osmoregulatory centre — a part of your brain next to the pituitary gland. This gland secretes the hormone ADH (anti-diuretic hormone). When you become dehydrated, your pituitary gland releases

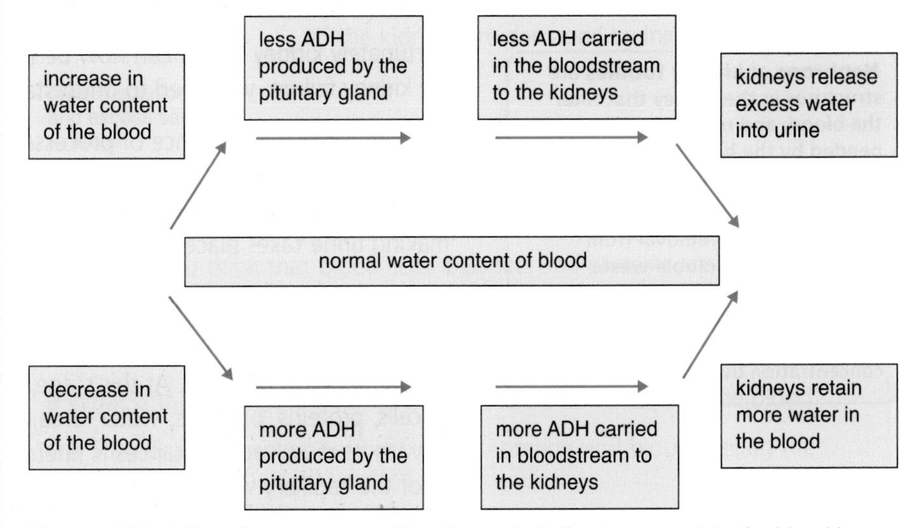

Figure 11.3 A flow diagram summarising the control of water content in the blood by ADH and the kidneys.

B3 11.3 Treating kidney failure

Learning outcomes

- Understand how kidney failure can be treated either by a dialysis machine or with a transplant.
- Be able to evaluate the advantages and disadvantages of treating kidney failure by dialysis or kidney transplant.

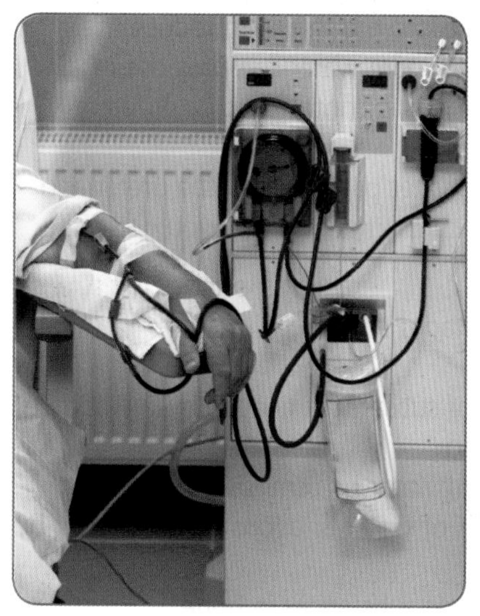

Figure 11.6 This patient has kidney failure and has to use a kidney dialysis machine regularly.

People may suffer from kidney failure for various reasons. Kidney failure can result from injury, infectious diseases or conditions that have been inherited. Kidney failure is treated in one of two ways — either by using a kidney dialysis machine or by a kidney transplant. There are a number of things to consider before a decision can be made about the type of treatment a patient will receive.

Kidney dialysis

Dialysis is used when the kidneys' function is at a level that would result in death if the patient were left untreated. Blood is taken from a vein in the patient's arm and fed into the machine. The blood is pumped along a tube made of partially permeable membrane. On the other side of the membrane is a liquid called dialysis fluid. This contains glucose and salts at the same concentration as the blood, but no urea. This means that useful substances (glucose and salts) are not lost from the blood, but harmful urea and any excess ions diffuse from the blood into the dialysis fluid through the partially permeable membrane. The treated blood is then returned to another blood vessel in the patient's arm.

Dialysis results in blood with the normal concentration of sugar and ions being returned to the patient's body. Someone using dialysis will normally use the machine three times a week for about 4 hours each time.

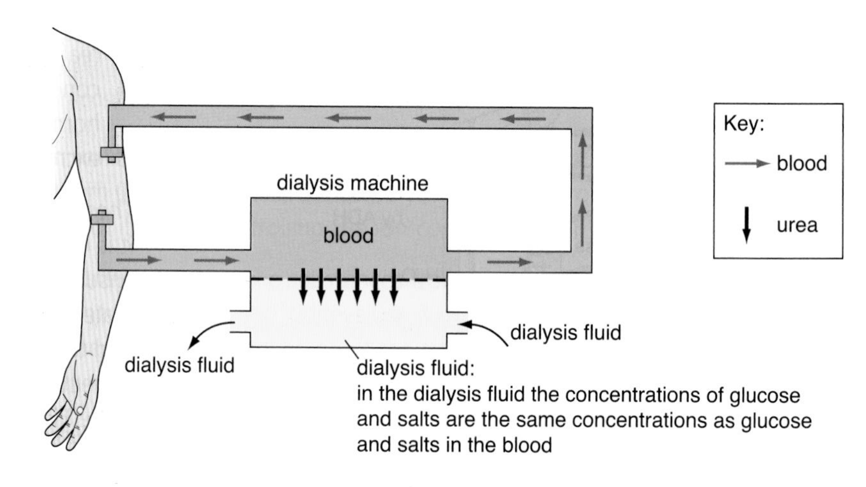

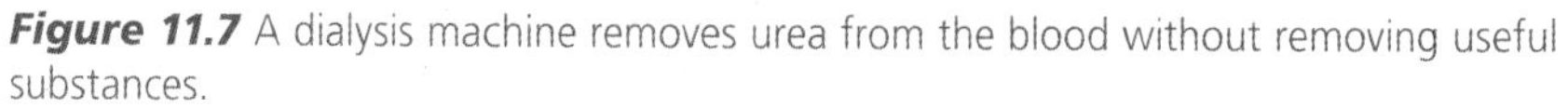

Figure 11.7 A dialysis machine removes urea from the blood without removing useful substances.

Test yourself

11 a For how long would a dialysis patient normally be attached to a dialysis machine during a week?
b How do you think this will affect the patient's lifestyle?

12 Blood entering the dialysis machine has a high concentration of urea. Explain how the urea is removed from the blood.

13 It is essential that the dialysis machine does not remove useful substances such as glucose. Explain how this is achieved.

14 How does the way that a dialysis machine work differ from that of a real kidney?

Kidney transplant

A kidney transplant involves a major operation to replace a damaged kidney with a healthy one from a donor. A donor may be a living person, often a close relative, who gives one of their healthy kidneys to the patient. This is possible because people can lead a perfectly healthy life with only one kidney. A donor may also be someone who has died after previously agreeing to donate body organs.

Look at www.uktransplant.org.uk/ukt/ and follow the links 'About transplants', 'Organ allocation', '... kidney donation' to see how the NHS decides who can receive a donated organ.

The operation takes about 2 hours and, if successful, replaces the need for dialysis. However, there may be complications. The biggest problem facing a patient who is judged to be suitable for a transplant is the lack of appropriate donors. This means that a patient may have to wait many months before an operation is possible.

What is rejection?

All cells have marker proteins on the surface called antigens. Before a transplant, the recipient and the donor will have had a 'tissue test' to verify that the two are compatible. It is very rare that donor and recipient are a perfect match, but a close 'tissue type' match means that fewer immune-suppressing drugs will be needed.

The patient's body may reject the donated kidney and the immune system may start to treat the organ as an infection and attack it with antibodies. If the kidney starts to show signs of being rejected, the patient is treated with drugs that suppress the immune system.

Test yourself

15 What are the advantages of a kidney transplant compared with treatment using a dialysis machine?

16 Apart from the medical problems, suggest one other problem that someone who has had a transplant may encounter.

17 When a kidney becomes available for transplant, there will always be more than one suitable recipient. In these cases, doctors must decide who will receive the organ. Working with a partner, make a list of factors that you think doctors should take into account. Then put your factors in order of importance.

How much does treatment cost?

Both kidney dialysis and kidney transplants are expensive. The approximate costs to the National Health Service are:

- £35 000 to keep a patient on dialysis for just one year, with dialysis being carried out at a hospital — there are about 21 000 patients on dialysis in the UK

Figure 11.13 Frederick Banting and Charles Best with the dog they used in their experiment.

Discovering insulin

HOW SCIENCE WORKS ACTIVITY

A doctor called Frederick Banting studied diabetes and thought that the condition was linked to the pancreas. With an assistant called Charles Best, he carried out a set of experiments using dogs. Best first tied up the tube connecting the dog's pancreas to its blood system. He then measured the dog's blood glucose levels and noted that these rose dramatically over an 8-week period. From his observations he deduced that the pancreas must produce something that controlled the amount of glucose in the blood.

Best then removed the dog's pancreas, along with the pancreas from another dog. He ground up the pancreas from the second dog with salt water, and injected this mixture into the first dog. Best observed that the first dog's blood glucose levels were temporarily decreased. However, the levels soon rose and the dog died the following day. Banting used these findings to conclude that the pancreas produced a substance that controlled blood glucose level — he called this substance insulin.

Banting went on to show that insulin obtained from healthy dogs could be used to treat diabetic dogs and won the Nobel prize for his work.

Once people realised that Banting's discovery could lead to a treatment for diabetes, further research into insulin was quickly carried out. It was found that insulin could be extracted from the pancreases of cows and pigs and be used to treat human diabetes successfully. Many thousands of lives were saved in this way, but a huge amount of insulin was needed, which meant that lots of pancreases had to be collected from abattoirs.

Nowadays, insulin is manufactured using genetically modified bacteria. This removes the need to use the pancreases from slaughtered animals (see 'Genetically engineered human insulin' in Section 4.6 and revise the benefits of Humulin).

1. Draw a flow diagram summarising the work of Banting and Best.
2. The first set of experiments were carried out using just two dogs. Bearing this in mind, do you think Banting was right to be so confident of his conclusion? Give reasons for your answer.
3. Suggest some improvements that Banting could have made to his experiment to collect more reproducible results.
4. People often refer to Banting's work when they argue for the use of animals in experiments to develop new medical treatments. Do you think that referring back to experiments carried out in 1922 is a strong enough argument to justify animal experiments today? Explain your opinion.
5. Some people objected to using insulin from animals' pancreases. What reasons do you think they had for this?
6. It was found that some people formed antibodies against the insulin from pigs. Suggest a reason for this.
7. The uses and methods of science and technology can raise ethical issues. Have an online debate with one group explaining and arguing for the work of Banting and Best, and the other group arguing that it is morally wrong to do this type of research on animals

Homework questions

1 a The girl in the photograph is losing water by sweating. How else might she lose water from her body?

b She is also losing dissolved ions in her sweat. Explain how this will affect the water uptake by her cells. You may need to look back over the section on osmosis in Section 9.1.

c How does sweating help her body to maintain the correct body temperature?

You can see that she is drinking to replace the water lost from her body.

d What would she have to do to replace the ions she has lost?

Imagine that the girl has only just started to dance.

e How would her pituitary gland respond when she initially starts losing water in her sweat?

f What effect would this have on her kidneys?

2 a Which part of the body monitors and controls body temperature?

b Describe, in as much detail as you can, how the body responds to an increase in body temperature.

c On cold days, many people appear to have very pale skin. Explain the scientific reason behind this observation.

3 a Which organ in the body does not function properly in somebody who suffers from type 1 diabetes?

b Explain how diabetes affects the body.

c Describe two ways in which type 2 diabetes can be treated in young people.

4 Healthy kidneys remove urea from the blood, but maintain the correct levels of sugar and salt. Explain how this takes place.

5 a If a kidney transplant were not a good tissue match, what could the consequence be?

b Why does a patient who has not had dialysis for 3 days feel a little under-the-weather?

c Explain why a sporty student may pass a small volume of dark yellow urine on a hot afternoon of athletics.

6 a Describe the role of the nephron or kidney tubule in removing urea and toxic substances.

b How does the body prevent loss of glucose via the kidneys when blood glucose level is normal?

c Explain how a hormone helps to control the body's water content.

d How will the composition of the blood differ in the renal artery and the renal vein?

Exam corner

a How is blood filtered in the kidney? *(2 marks)*

b Explain how substances needed by the body are reabsorbed in the nephron. *(4 marks)*

Student A

a The blood enters the outer part of the kidney at high pressure ✓ and all the dissolved substances ✓ are forced through the gaps in the capillary wall into the nephron.

b The nephron is a long, thin tube. Glucose is reabsorbed, water is reabsorbed, ions are reabsorbed but urine ✗ is not reabsorbed.

Examiner comment In part (a) Student A has the right idea and scores both marks.

Part (b) asks *how*? Student A has given a list of substances but has not explained how. A second mistake, which is commonly made, is to confuse urea and urine — be aware of this.

Student B

a The blood is filtered in the outer part of the kidney by pressure filtration ✓ and the blood ✗ goes into the kidney tubule. ✗

b The water is reabsorbed by osmosis ✓; the sugar is reabsorbed — firstly by diffusion and then by active transport ✓, some ions are reabsorbed but urea is not and it passes to be collected in the bladder. ✓

Examiner comment Student B's answer to part (a) is not exact enough because it implies that the plasma and cells etc. pass into the nephron. Pressure filtration separates the plasma and dissolved substances from the blood cells and platelets.

Student B gives a good answer for part (b).

Neither student has stated that water reabsorption is controlled by the hormone ADH. Student B could have expanded the statement about water reabsorption by stating that if the body is dehydrated, more ADH is released from the pituitary and so more water is reabsorbed through the walls of the collecting duct.

Chapter 12 Humans and their environment

Setting the scene

Solar CooKit and WAPI save the lives of many children. Water pasteurisation using heat from the Sun and a reflector is the sort of technology that you wonder why it wasn't thought of before. Can we make more use of renewable energy sources and cause less damage to the environment?

Practical work

In this chapter you can learn to:

- decide which variables to control when investigating compost
- identify the equipment needed for a model biogas generator

ICT

In this chapter you can learn to:

- use a range of media to present information about ecosystems in your area
- select the most effective way to present information on human impact on your local environment
- create online debates about logging and global warming
- use Facebook to get across ideas on improving the environment to a wide range of people

B3 12.1

How do humans affect the environment?

Learning outcomes

- Understand the consequences of an increasing human population on the depletion of raw materials.
- Be aware of the reasons for an increasing area of land use by humans.
- Be aware of increasing production of waste and pollutants by humans.

There are about 6.9 billion people in the world today and the number is increasing. How can the requirements of this increasing human population be met without damage to the environment? People in the UK expect to have a house or a flat to live in, and have easy road access to supermarkets and other shops. In their homes they want electrical goods such as a fridge, freezer and washing machine, and many have a computer and media system. Can all these be provided without any impact on the environment?

Does an increasing human population reduce the amount of land available for animals and plants?

An increasing human population needs more land for:

- building homes, industrial estates, motorways and retail parks
- quarrying and mining of raw materials for buildings and roads
- intensive farming
- waste disposal

Currently between 10 and 20% of the UK's native species are threatened with extinction by expanding towns, motorways and intensive agriculture. Sites of Special Scientific Interest (SSSIs) have been established in some areas to support a rare species or group of species. Such species need special protection and the **ecosystem** around them needs to be large enough to remain balanced and undisturbed.

An **ecosystem** is made up of the plants, animals and microorganisms (the community) living and surviving in one place and their interaction with the surrounding non-living environment.

A balanced ecosystem will change slowly over time — this means that the number of plants and animals and the range of species remain similar from one year to the next. Size is very important when setting up an SSSI. Animals and plants need a certain area to sustain a breeding population. Dividing up an area of land, for example by putting a motorway through it or golf course on it, may result in the separate parts suddenly becoming too small to feed populations of larger animals such as deer or badgers.

Figure 12.1 A road-widening scheme carves through woodland, dividing the ecosystem.

Having examined the consequences of making an ecosystem smaller, think about the opposite case of making it larger. Removing hedges and ditches to make bigger fields removes the shelter, feeding and breeding sites for animals, including biological control species, such as predatory beetles, insectivorous birds and hedgehogs. These are the carnivores in the food web. By eating pests such as greenfly and caterpillars, these predators can keep pest numbers low naturally. Removing hedges and ditches upsets the balance of an ecosystem, disrupting food webs and allowing pests to reproduce and cause further crop damage.

Ecosystems HOW SCIENCE WORKS ICT

Create a multimedia resource for Natural England on the state of ecosystems in your area. Are there any SSSIs or should there be some? What could be done to improve the area for wildlife? You can do this as a presentation, podcast or video.

Test yourself

1 An increasing population leads to increased transport. List three ways in which this can affect ecosystems.

2 Modern farming with big machinery means big fields. What are the ecological consequences of this?

Does an increasing human population have an impact on raw materials?

Important raw materials — such as building stone, metal ores and fossil fuels — are finite resources. Their increased use means that they will run out more quickly and alternatives — such as renewable energy sources and recycled building materials — must be developed. Trees take many years to grow to a size where the wood can be used, so regular replanting must take place to prepare for future needs.

Can the volume of waste deposited in landfill sites be reduced?

Low-lying areas of land and old quarry pits are commonly used as landfill sites. They are filled with waste, which is then compressed. Although these sites are well regulated there are still environmental problems associated with them — sites can be unsightly; birds that feed on the waste, such as gulls, can transport pathogens; and toxic chemicals can be washed into waterways after heavy rainfall.

Figure 12.2 NIMBY?

Landfill sites are not an ideal solution to the problem of disposing of our rubbish. Many sites around the UK are now almost full, and in some cases no alternative site is available locally.

Waste disposal is a complex problem. New solutions to the increasing problem need to be found.

Composting

HOW SCIENCE WORKS PRACTICAL SKILLS

Research and build a compost bin. What types of waste can be composted? What are the ideal conditions for producing compost? How would composting waste help to preserve peat bogs?

Test yourself

3 What problems are associated with landfill?

4 What are the advantages of recycling?

Figure 12.3 Do you know where your nearest recycling facility is located?

Currently:

- the UK produces over 400 million tonnes of waste per year (you throw away 10 times your body weight of rubbish every year)
- the increasing population of the UK and higher standards of living mean that waste production is expected to rise
- a large percentage of rubbish is still disposed of in landfill sites — but this amount has been reduced and the government has set targets for further reduction
- a considerable proportion of the material previously placed in landfill sites is now being recycled to save raw materials

So what can be done with all our rubbish? Alternatives to landfill include recycling, incineration and composting. All of these alternatives need processing plants, which cost money to build — although some of the costs could be met by the sale of materials.

Recycling

This involves the mechanical sorting of mixed rubbish — for example, pulling iron and steel objects out of a mixture using a giant electromagnet — and manual sorting by each household into separate bins for glass, tins, plastics etc.

Out of sight, out of mind

HOW SCIENCE WORKS ACTIVITY

It is easy to throw rubbish into a bin and forget about it.

Find out about each of the following waste disposal methods — recycling, incineration and composting. Construct a table of advantages and disadvantages for each.

You can use the internet to help you to do this. Search for 'waste' on the UK Environment Agency website, www.environment-agency.gov.uk. Type in your postcode on this website and find out how rubbish is dealt with in your area.

You may be surprised by how far rubbish travels — for example, waste paper collected on the south coast is taken to Shotton on Deeside for recycling for printing newspapers.

Recycling

HOW SCIENCE WORKS ACTIVITY

In a group, examine your school's current recycling efforts and make suggestions for improvement.

Incineration

This process further reduces the volume of rubbish placed in landfill sites. The ash produced can be used for road building, saving natural stone. All new incinerators are also 'EfW' (energy from waste) plants. This means that the heat produced by burning the rubbish is used to make electricity. Proposals for building new incinerators frequently encounter opposition from local residents who fear the risk of air pollution. Modern incinerators are highly sophisticated

Biodegradable material is usually of plant or animal origin. It can be decomposed by bacteria or fungi. Paper, cotton and wool waste are all biodegradable.

Fig 12.4 Reed filtration bed — this results in a new and interesting ecosystem.

Test yourself

5 Composting reduces biodegradable rubbish. What gas is produced in landfill if biodegradable rubbish is tipped? What problem does this cause?

6 Water meadows are grazed but are not treated with fertilisers or herbicides. Why not?

Boost your grade

Do not confuse fertilisers and herbicides. Fertilisers are plant food (remember 'f' and 'f') used to make the plants grow bigger and faster. Herbicides are used to kill plants that are regarded as weeds ('-cide' means 'to kill', as in homicide). Herbicides are used to reduce competition between the weeds and the crop for nutrients, light and water and this increases the yield of crops.

industrial plants where all emissions are released at safe levels and are monitored by a computer 24 hours a day. They do not pose a risk of releasing sulfur dioxide and causing acid rain because the waste gas passes through a desulfurisation process, forming a solid sulfate that can be removed.

Composting

This is used to reduce the volume of **biodegradable** materials. After treatment (to kill pathogens) the compost can be used or sold and can be used as a substitute for peat.

What is the impact of an increasing population on waste-water treatment?

Before the Industrial Revolution, people lived in small, scattered rural communities. Waste water was emptied into ditches and the biodegradable contents decomposed naturally. Nowadays towns have much larger populations. If sewage was emptied into our rivers it would create an awful smell, and fish and other wildlife would be poisoned. The treatment of waste water and sewage has improved to meet the demands of an increasing population. Today only treated sewage passes into rivers or out to sea and it must not contain pathogens or toxins (poisons).

Many pharmaceutical companies and power stations use a biological solution to overcome waste-water problems — they filter waste water through reed beds. The reeds use the nutrients in the waste material to grow, and at the same time remove toxins from the water. Reed beds provide an interesting habitat for wildlife. Water run-off from motorways and some airports, such as Heathrow, is treated by flowing through such reed beds.

Intensive farming involves using soluble fertilisers, pesticides and herbicides to increase crop production. If fertilisers wash into rivers they cause the algae and plants to grow vigorously — this has two effects. Firstly, the vegetation blocks the light so that plants below the surface cannot photosynthesise and oxygenate the water. Secondly, because of deoxygenation, fish, plants and invertebrates die and the river becomes 'dead', the water turns brown and smelly and is unfit for anything. To overcome this, areas around rivers and water-catchment areas are not fertilised and are used mainly for grazing. Toxic chemicals, such as pesticides and herbicides, are not used on grazing land and so do they not get into the water.

Human impact on the local environment — HOW SCIENCE WORKS ICT

Create an online blog to debate the human impact on the environment in your local area. What damage has been caused to the environment? What is being done to reduce this damage? What more could be done?

This section started with the question 'How do humans affect the environment?' After examining four different issues, we can now make some positive suggestions. As a society, we must:

- avoid changing balanced ecosystems because this can have serious consequences and damaging long-term effects

The term **brown-field site** is used to describe an area in a town where old properties have been demolished. A **green-field site** is farmland or green belt land surrounding a town where there have been no buildings before.

- increase recycling and the reuse of materials such as metals, plastics, glass and paper — this will save raw materials
- reuse **brown-field sites** in towns for new housing — this will save farmland and new roads will not be needed
- use biological methods, where possible, to treat waste water and biodegradable waste — the end products are harmless to the environment

B3 12.2

Learning outcomes

- Understand how the removal of rain forests and peat bogs results in an increased release of carbon dioxide.
- Appreciate that the loss of forests results in reduced biodiversity.
- Know why deforested land quickly becomes too poor for agriculture.
- Understand the conflict between the economic development and the environment.

Deforestation and the destruction of peat areas

Deforestation

Large-scale deforestation of tropical rainforests throughout the world has been, and still is, widely publicised. The Amazon rainforest is a major source of global oxygen and many pharmaceutical drugs have been developed from rainforest plants. We must take steps to ensure the survival of this resource. It also shows that the problems associated with an increasing worldwide population are not restricted to developed countries.

The ecosystem of a rainforest depends on the recycling of nutrients. Dead leaves are decomposed by bacteria and fungi, and the minerals are released into the soil. 'Slash and burn' also releases minerals into the soil, and these can support the production of arable crops for a small population for a few years. As the minerals in the soil are removed and not replaced, crop yields fall and the village population moves on, abandoning the clearing. The natural vegetation will regenerate from seeds in the surrounding forest, but this takes time. Very different effects result from commercial logging as can be seen from Figures 12.5 and 12.6.

Figure 12.5 When this village moves on, the area will recolonise from seeds blown in from the surrounding forest.

Figure 12.6 Vast tracts of deforested land do not regenerate.

Deforestation provides:

- timber that is sold for money
- timber for use as a building material
- land for cattle ranching
- land for growing crops to generate income — e.g. bioethanol
- land to build roads which connect major cities

Large-scale logging results in:

- vast areas being cleared — this affects the rainfall pattern
- removal of the tree canopy, which breaks heavy rainfall — this causes soil erosion as water from heavy rainfall flows across the land
- compaction of the soil and root damage by heavy machinery, preventing drainage and causing flooding
- flooding when soil washes into streams and rivers and causes a blockage
- huge areas of deforested land being abandoned because the thin, acid soil cannot support crops or cattle — these areas are becoming deserts
- isolation of animal populations, which will not cross large open areas where forest has been cleared
- reduction in the sources of food for both human and animal populations
- loss of biodiversity — a reduced number of plants species, some of which are of medicinal value, and reduced numbers of animal species

Scientists from the Massachusetts Institute of Technology gathered evidence by examining 75 years of rain-gauge records. These showed a big increase in rainfall over large deforested areas. The scientists also examined current satellite data about cloud cover. Satellite pictures showed twice as much low-level rain-bearing cloud over large deforested areas, compared with forested areas. Small deforested areas do not give the temperature differences needed for the formation of clouds.

The Brazil government uses satellite surveillance technology to help in preventing illegal logging. It faces the dilemma of trying to improve the lives of the increasing Brazilian population while at the same time preserving the rainforest.

A rainforest appears lush but it is still a delicately balanced ecosystem. In an undisturbed ecosystem, the carbon dioxide uptake for photosynthesis is balanced by that produced by respiration and decomposition. Logging results in the release of higher levels of carbon dioxide by:

- the increased activity of soil microorganisms, as leaves and branches from felled trees decompose
- the burning of waste wood in fires

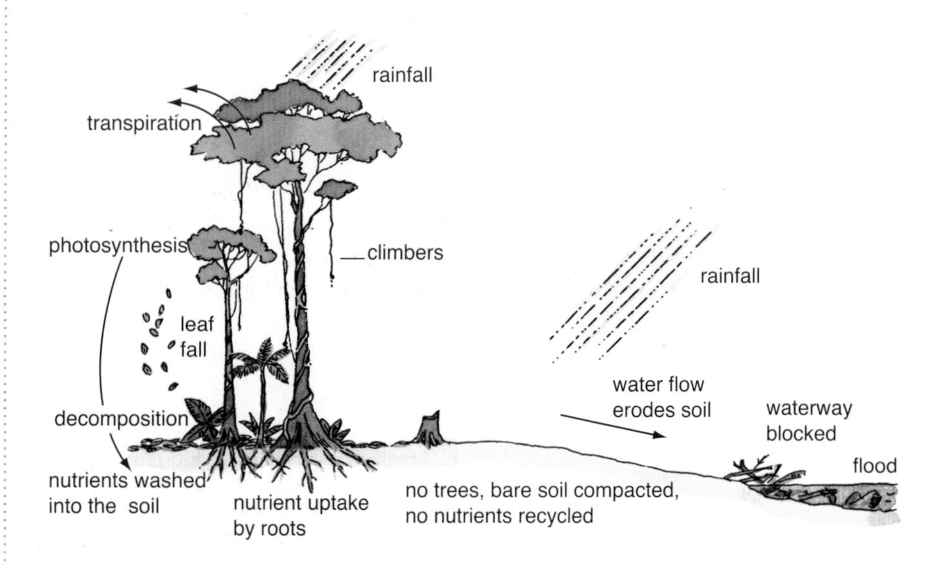

Figure 12.7 The rainforest cycle and biological consequences of disturbance.

Test yourself

7 a Who do you think provides the evidence for the regrowth of vegetation in small cleared areas of rainforest?

b What type of evidence would you need to be convinced that regeneration takes place? Briefly describe a project to investigate this.

c What is it about the soil in a rainforest that limits its value as agricultural land?

d Why does soil erosion occur in deforested areas?

e (i) What did scientists from the Massachusetts Institute of Technology discover?
(ii) Do you think their evidence can be reliably linked to flood damage reported in vast areas cleared of forest?

f List what you see as the harmful long-term consequences of extensive deforestation of rainforest. Will the effects be local, global or both?

g Using your answer to question **f**, suggest a management strategy that would result in sustainable development of the rainforest.

been a great deal of debate about global warming, it now appears that scientists have **underestimated** the rate of the rise in world (global) temperatures.

The consequences of global warming are that a rise in temperature of only a few degrees Celsius is enough to change the climate. This could mean a change in the direction of rain-bearing winds, resulting in heavier rainfall in some parts of the world and severe drought elsewhere — this will reduce biodiversity. Crops that can grow will change as a result of drought. If the melting of ice continues there will be a further rise in sea level and areas of Holland, eastern England and Bangladesh that are just above sea level could be flooded.

B3 12.4

Learning outcomes

- Appreciate that waste must be handled correctly from the source.
- Understand that we all have a role to play in conserving finite resources.

Reducing human impact on the environment

Having collected and analysed evidence from measured data and scientific models, we can start to form conclusions. All governments should be concerned about global warming and they should take action to reduce their emissions of carbon dioxide and methane. Individuals, local government and industry can all make changes, based on scientific evidence, to reduce the negative human impact on the environment.

People who do recycle materials are acting responsibly to conserve natural resources and to reduce landfill. We all have a part to play in meeting the targets of local and national schemes to reduce waste. Local authorities are making recycling easier for us. Not only are there bottle banks and facilities for recycling aluminium cans, but many towns now also have kerbside collections for newspapers, bottles and cans. Countries such as Germany have been doing this for longer than the UK and have reduced waste significantly.

Figure 12.9 More and more young people are actively recycling and trying to reduce waste. Is there a point to this?

Making our homes and offices more energy efficient by installing loft insulation, cavity-wall insulation and double-glazing reduces the amount of fuel needed for heating. When it gets chilly, for example, people could wear a fleece or jumper indoors instead of turning the heating up. These changes are beneficial because:

Figure 12.10 Where is your nearest cycle route? Is there a safe cycle route to school or to a leisure centre?

- fossil fuels will last longer
- greenhouse gas emissions will be reduced
- domestic gas, oil or electricity bills will be reduced

Walking or cycling instead of driving improves health, as well as:

- saving fuel costs for private vehicles
- reducing air pollution
- saving petrol and oil, which are non-renewable resources

Public transport vehicles, which can carry many people, provide a similar set of benefits to the environment. Some government officers are using public transport or walking instead of being taken in ministerial cars.

Improving the environment

HOW SCIENCE WORKS **ICT**

Design a Facebook group for improving your local environment.

What information and pictures will you put on it?

What can you do to make a better future?

HOW SCIENCE WORKS **ACTIVITY**

1. If your town has kerbside collections, find out which materials are collected separately. Look up 'waste local plan' or look on your local council's website.
2. Make a list of common products that can be produced from recycled materials.
   You may find www.recycledproducts.org.uk and www.sustainable-development.gov.uk useful for some ideas for your research. Alternatively, you can log onto the RSPB website www.rspb.org.uk and search for 'green living'.
3. **Planning**
   In a small group, develop an environmental plan that involves avoiding waste or increasing recycling. Try to think of a new idea that supports the plan that you could put into action in either your school or the community in which you live. For example, a local charity might have a fund-raising recycling project for which you could provide support. On a street map, mark the neighbourhood recycling facilities.
4. **Communicating**
   Create a poster, webpage or PowerPoint presentation that outlines your plan to others in your class or school. Remember to explain the importance of each suggestion in your plan. For example, recycling aluminium cans saves both the raw material and the energy used to refine aluminium ore. Plastic bottles and paper are easily collected and recycled.
5. **Action**
   Obtain permission to display your posters etc. Include details of local recycling facilities or write to the school Head asking to provide recycling facilities on site.

Biogas is a mixture of gases produced when bacteria break down plant or animal waste. It is used as a heating fuel or to power electricity generators.

Anaerobic fermentation is fermentation without oxygen. Fermentation is a type of anaerobic respiration by microorganisms. The reaction produces gases and other products from the breakdown of carbohydrates.

intestines of animals and are found in animal dung. These bacteria function most efficiently at 30–40°C and in anaerobic conditions. However, at the start of the process there is air in the tank, and so the first gas produced is carbon dioxide. This is the product of decomposer bacteria respiring aerobically, as in an open compost heap. When most of the oxygen has been used up, a mixture of approximately 60% methane and 40% carbon dioxide is produced. Because the bacteria produce methane as well as carbon dioxide, they are described as methanogenic.

Biogas generators

Compare these two biogas generators.

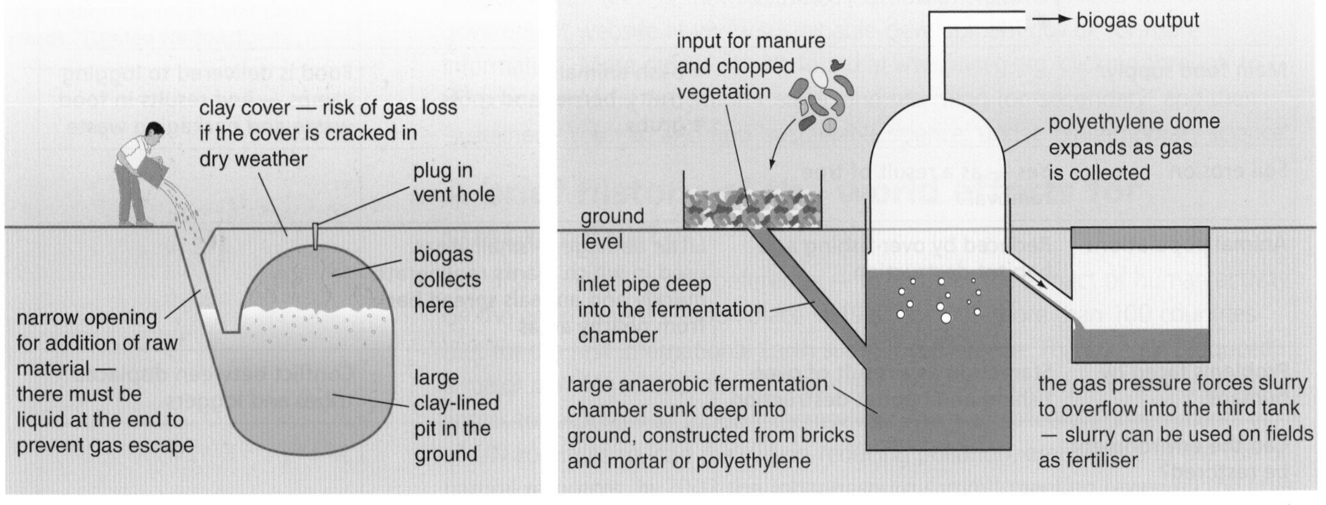

Fig 12.12a Simple rural biogas generator — this is a batch process. If production rate drops the tank has to be emptied and cleaned out, and the cover remade. There will be a gap in production.

Fig 12.12b Modern biogas generator — this is a semi-continuous process. The manure from two dairy cows can provide a fairly continuous supply of biogas to meet the cooking and lighting needs of a small family.

Evaluating biogas generators

How do biogas generators work? Each type of bacterium in a biogas generator functions under its own optimum conditions.

In a biogas generator, supplied with farm waste such as chopped straw and cow dung, bacteria in the first stage of the process must produce enzymes that break down cellulose to sugars and finally glucose.

Figure 12.13 This is a small version of Figure 12.12a. Can you identify the parts?

In the next stage, other bacteria produce enzymes that convert the sugars to organic acids.

The final group, methanogenic bacteria, produce enzymes to convert these compounds to methane. If there are proteins in the farm waste these are broken down by bacteria that produce protein-digesting enzymes. Methanogenic bacteria cannot function at very low pH values. However, the fall in pH caused by the organic acids is neutralised by ammonia made during protein breakdown, so the methanogenic bacteria can survive. Gas production also depends on temperature. The bacteria function best at about 35°C — so biogas is not as easy to make in northern Europe as in hot countries. Fortunately, the fermentation reaction in the generator is exothermic, and the generator can be insulated to keep this heat in.

Figure 12.14 Is this an improvement on a wood fire?

Next we examine the social and health advantages. Before biogas was available, the main fuels used in developing countries for cooking and heating were wood and dried cattle dung. These fuels have several disadvantages, compared with biogas:

- The wood for fuel is taken from forests, causing trees to be damaged or lost. Over time, loss of trees causes soil erosion, making it difficult to grow other plants. Biogas uses waste materials rather than wood and so allows trees and scrubland to regenerate.
- To collect wood for fuel, people need to make long, exhausting journeys carrying heavy bundles of wood. Biogas uses materials that are closer to hand.
- Burning wood inside a poorly ventilated home produces smoke and fine particles that can damage lungs and cause health problems. Cooking on biogas stoves is clean and odourless.
- Using dried cattle dung as a fuel means there is less manure to improve the soil, and so crop production decreases. While biogas can also be produced using dung and manure, the waste from the biogas generator can be used as a fertiliser, increasing crop productivity.

Biogas, like wood, is formed from a renewable source — it will never run out. Biogas is also a **carbon-neutral** fuel that does not contribute to increasing carbon dioxide emissions. This is because, even though carbon dioxide is produced in the decomposition process and again when biogas is burned, the carbon dioxide was removed from the air when the original plant material was growing.

Test yourself

8 Name the first gas produced in a biogas generator, just after the waste material has been added. Is this gas combustible?

A carbon-neutral fuel causes no overall increase in carbon dioxide levels in the atmosphere when it is burned, because the carbon dioxide taken up during photosynthesis equals that produced during combustion.

Biogas comes to Indian towns

HOW SCIENCE WORKS ACTIVITY

Biogas generators are now common in rural areas of India where there are cattle providing lots of manure. Unfortunately the conventional type of generator cannot be used in towns and cities, where there is no livestock. However, Dr Anand Karve from the Appropriate Rural Technology Institute (ARTI) in India invented a small generator that could be used in urban areas. In 2006, he won the Ashden Award for renewable energy.

A conventional biogas generator, sunk into the ground, works on human or animal excreta and starts producing sufficient gas after 40 days. Dr Karve's new compact generator can use any starchy, cellulose-rich or sugary vegetable food waste, and gas production starts after only 2 days. What's more, this generator produces 1.0 m^3 of biogas from only 2 kg of waste, compared with the 80 kg of manure needed to produce 1.0 m^3 of biogas in a traditional generator. 2 kg of waste is easily achieved with fruit and vegetable skins, spoilt milk and food leftovers and this produces enough gas for cooking each day by the average urban family.

The gas produced is carried from the generator, situated just outside the house, directly into the kitchen biogas stove. A biogas generator replaces the cylinders of 'bottled' gas, which are expensive, heavy and need collecting or delivering.

Figure 12.15 A proud owner standing next to his urban biogas generator. The generator is small enough to fit in an urban garden or terrace.

1. How does the structure of the new biogas generator compare with the conventional rural type?
2. What are the advantages of the new compact biogas generator for urban populations?
3. Why can the urban biogas generator be smaller?
4. In addition to the production of biogas for cooking, what is the main environmental advantage of the new generator?
5. a State the typical composition of biogas.
   b Assuming a 60 : 40 composition ($CH_4 : CO_2$), what volume of methane will be present in 500 dm^3 of biogas?
   c Why do you think the energy produced from biogas can vary?
   d Why do you think a typical cooker has an efficiency of only about 60% when using biogas compared with the same cooker using methane?
6. The fuel that biogas replaces in urban areas is 'bottled gas', or methane. The energy value of methane is 36 120 kJ/m^3.

   The energy values of biogas are typically:
   - 21 500 kJ/m^3 for 60% methane
   - 25 200 kJ/m^3 for 70% methane

   a Using these figures, copy and complete Table 12.2 to calculate the volume of gas required to boil 1 dm^3 of water using two biogas fuels that have different proportions of methane. Assume that the starting temperature of the water was 20°C and that 4.2 kJ raises the temperature of 1 dm^3 of water by 1°C.

	Biogas with 60% methane	Biogas with 70% methane
Temperature through which the water has to be raised in °C		
Energy required to raise 1 dm^3 water to its boiling point in kJ		
Volume of gas required = energy required in kJ/energy value of the fuel		

Table 12.2 A comparison of biogas fuels.

   b Why will a family benefit from including waste with a higher carbon content?
7. A typical biogas cooker uses 0.3 m^3 of fuel per hour and requires 2 kg of feedstock to produce 1.0 m^3 of biogas.

   What mass of feedstock is required to provide sufficient gas for 3 hours of cooking?
8. Suggest two factors that will affect the efficiency of a biogas stove.

Figure 12.16 Solar CooKit.

Solar reflector stoves

Even more environmental and health benefits come from solar reflector stoves that can be used for water pasteurisation (WAPI) and save many children from gastric infections caused by pathogens. They are a great advantage in refugee camps such as Kakuma Refugee Camp — home to about 100 000 refugees. This part of Kenya is extremely dry and wood is very scarce. A web search for 'Ashden Awards' will provide more information on this and other projects.

Using large-scale biogas generators for waste disposal

Large biogas generators can solve the waste-disposal problems of intensive pig and poultry farms. Pigs produce a large volume of waste, which must not be allowed to pollute water. The problem is severe in countries such as Holland and Denmark where there is intensive pig production but not enough farmland on which to spray all the manure produced. Other problems can result if heavy rainfall washes the manure into rivers, where it can cause severe environmental damage by making algae and plants grow rapidly. As the algae die, aerobic bacteria decompose them and the water becomes brown, anaerobic and very smelly.

The large biogas generator shown in Figure 12.17 operates almost entirely on pig slurry. This is collected in tanks before being fed into the generator. The generator is kept at an optimum temperature by burning some of the biogas produced. The biogas can also be burned to dry the final product to make a useful fertiliser. Gas production is large enough to heat the farm buildings and to power an electricity generator for the local village.

Make a model generator

HOW SCIENCE WORKS
PRACTICAL SKILLS

Research and make a model biogas generator from scrap such as an old plastic drink bottle. Consider how you will collect the gas, insulate the generator and what waste you might use.

Figure 12.17 This biogas plant produces enough biogas to generate electricity for a village in the Netherlands. The circular slurry tank contains raw manure that is fed into the large anaerobic digester.

Test yourself

9 State one economic advantage of using biogas.

10 State four ways in which the health and lifestyle of people in developing countries have improved as a result of using biogas.

11 What are the benefits to the environment of using biogas as a fuel in countries such as Nepal?

12 Why must the final product from a refuse biodigester be pasteurised before it is used on greenhouse crops?

13 If a biogas digester uses restaurant waste as a raw material, suggest what enzymes would need to be produced by the bacteria for digestion of the material.

14 What are the environmental and economic benefits of using a biogas generator for intensive pig farms?

15 How would the conditions in a biogas generator in Scotland be kept at optimum levels in the winter?

Mycoprotein is formed from the mycelium of a fungus — a mycelium is the mass of fine fungal threads called 'hyphae'.

Quorn is made from **mycoprotein** produced from the fungus *Fusarium*, which is found in the soil.

The quorn is made in a continuous-flow process. The fermenter is a tall stainless steel container (about 40 m in height). The inside is kept sterile. A computer controls conditions so that they are optimal for the growth of *Fusarium*, with the temperature maintained at 32°C and the pH at 6. The sterilised nutrients are glucose (formed by breaking down starch using enzymes), mineral salts and vitamins. Sterilised air with some ammonia is pumped into the fermenter. The ammonia is a source of nitrogen for making proteins with the glucose. The conversion of glucose to protein is fast, and it takes only 5 hours for the biomass to double. A continuous feed of nutrients and the simultaneous removal of broth containing the fungal threads results in a continuous fermentation process. The fungal threads are then collected using a centrifuge and dried. After this, shaping and flavouring can take place.

Figure 12.19 Quorn is an edible substance made from mycoprotein. What is mycoprotein?

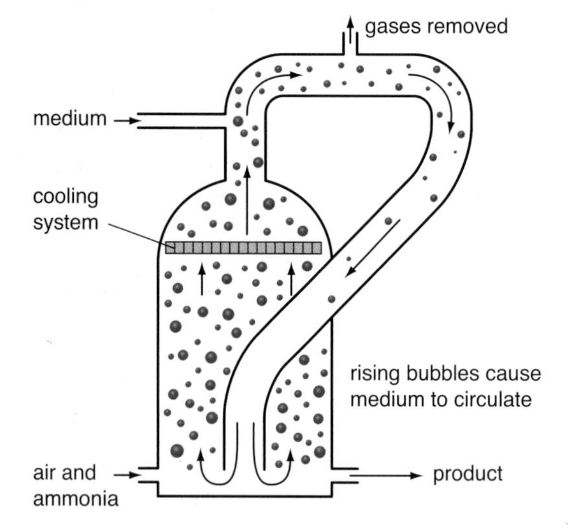

Figure 12.20 A fermenter for the production of mycoprotein.

In the production of mycoprotein, it is essential that pure cultures of the microorganism are used and that no unwanted microorganisms enter the fermenter. So, the tanks go through cycles of steam sterilisation after each production cycle. All nutrients must be sterilised and the air bubbled in must be microfiltered to remove bacteria and/or fungal spores.

Test yourself

17 Mineral salts added to the mycoprotein fermenter include potassium, magnesium, phosphates and trace elements. What is the value to the consumer of adding these:
  a to *Fusarium*?
  b to Quorn?

18 What use do you think *Fusarium* makes of the ammonia pumped into the fermenter?

19 The process for the production of mycoprotein is described as a continuous-flow process, compared with the batch process for penicillin. Explain what this means.

An alternative to meat

HOW SCIENCE WORKS
ICT

Create a marketing campaign for Quorn, basing it on the environmental benefits of eating it rather than eating meat.

Sustainable fish production

Sustainable fishing practices should mean that the population can be maintained indefinitely — in other words, allow fish to grow to reproductive size and to breed before they are caught. This should allow the UK fish population to increase and prevent fishing to extinction. Sustainable fishing will not adversely impact on other species within the ecosystem by removing their food source, accidentally killing them or damaging their physical environment. A UN report suggests that more than 70% of fishing grounds are 'over-fished' or that 'stocks are seriously depleted'.

The exact method by which the fish are caught is very important. Bottom trawling (such as for plaice) drags a huge weighted net over the seabed (Figure 12.21). This damages the seabed ecosystem and scoops up many different species.

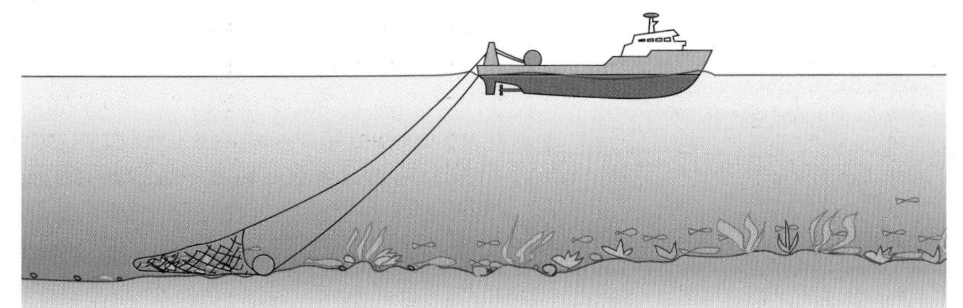

Fig 12.21 Bottom trawling for plaice. What damage do the rollers do to the ecosystem?

Pair trawling (such as for cod) uses two boats to drag a wide net (Figure 12.22), but this catches many unwanted species of fish and other animals, such as dolphins, albatrosses and turtles. These are thrown back, but many are dead by this time.

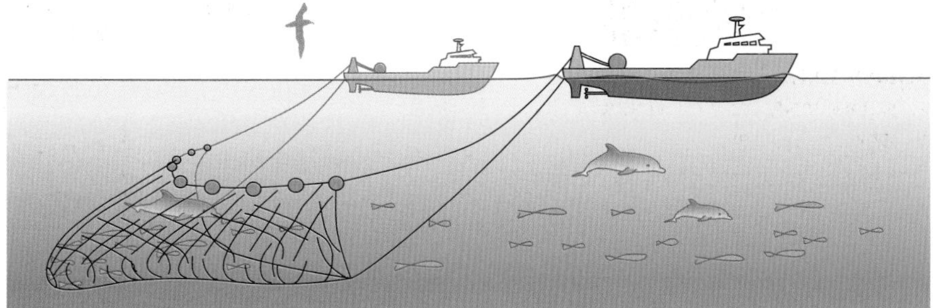

Fig 12.22 Pair trawling for cod. What else gets caught in the wide net?

The size of the net mesh is critical — if large enough, any small fish (not big enough to eat) will be able to escape and to carry on growing. Populations of some fish — such as cod, haddock and whiting — are so low that they need areas left unfished in order to breed.

To allow fish populations to increase the Common Fisheries Policy set total allowable catches (or fishing quotas) which fix the total amount of fish that can be landed by fleets in each one of the EU member states. This has a severe flaw — the catch is sorted at sea and only the best fish are kept, the rest being discarded, and usually dead by this time. If fish populations do decline too far and there are not enough full-sized adults left to breed then recovery will take longer, and the catches of trawlers will decrease. Species that used to be commonly found may even cease to exist in some spawning (breeding) areas.

An alternative method of fish production used is fish farming. This is growing at a rate of 8% per year worldwide. In the UK there are over 500 fish farms producing in excess of 8000 tons of fish, as well as shellfish farms. Because this is intensive culturing, there are welfare problems such as physical damage, disease and parasites to be controlled. There also needs to be a water current to flush the nitrogenous waste from the farm. Also, escaped fish may interbreed and weaken the wild stock.

Figure 12.23 A salmon farm showing high fish density. What problems does this cause?

The distance travelled by land or air from food producer to consumer is called **food-miles**.

Foot-and-mouth disease is an infectious disease affecting cattle, sheep, pigs, goats and deer. The sale of livestock is severely restricted by an outbreak of the disease.

BSE ('mad-cow disease') is a neurological disease that affects cattle. On 8 March 2006, the export restrictions against export of British beef to Europe were lifted — the feed controls and inspections had been effective in controlling the disease.

Herbicides are chemicals used to kill weeds so the weeds do not compete with the crops for water, light or nutrients.

Insecticides are used to kill insect pest species that eat and damage crop plants — but insecticides can also kill beneficial insects, such as bees.

Biodiversity is the number of different types of plants and animals living and feeding in an ecosystem. These may include pollinators and biological control species that are beneficial to crops.

The hidden costs of food production and distribution

The following points highlight some of the conflicts between the demands of shoppers for a good quality product at the lowest possible price, and the needs of the producer to increase productivity and reduce costs:

- Crops sold in supermarkets are produced intensively using fertilisers, irrigation, and, if not organic, herbicides and insecticides as well.
- Where greenhouses are used, there will be extra costs for the input of heat, light and carbon dioxide.
- Exotic or out-of-season crops that are grown abroad have to be flown or shipped to the UK at extra cost, clocking-up **food-miles**. Transport uses up non-renewable fuels. People living in rural areas have the benefit of being able to purchase local produce direct from farm shops.
- Meat production costs include those of rearing, feed material, housing, waste disposal, transport to market, veterinary bills and labour.
- Distribution can also result in spread of infection. Serious outbreaks of **foot-and-mouth disease** informed the public of the distance travelled by livestock from rearing locations to fattening locations, often hundreds of miles away.
- **BSE** made the public aware of the risks of intensive rearing which, in the 1980s, used animal carcass protein as a protein supplement in food — this practice is now banned.
- There is concern about the spread of bird flu in intensive poultry units, and poultry imports to the UK from certain countries are banned.
- Conflicts exist between agricultural production methods such as the use of fertilisers, **herbicides** and **insecticides** and damage to the environment — for example excess weed growth in rivers and streams and loss of **biodiversity**.

Food production issues

HOW SCIENCE WORKS ACTIVITY

Are shoppers becoming more aware of food production issues?

Get a large piece of paper (A3). In a small group, using a pencil and ruler, mark the paper out into four sections, as shown below.

Food production	Food distribution
Positive aspects	Positive aspects
Negative aspects	Negative aspects

Write down the positive and negative effects of managing food production and food distribution. For example, you might say that the use of an insecticide made sure that there were no holes in the cabbage leaves and no caterpillars lurking between the leaves!

You might like to include other topics such as selling only seasonal or local produce, genetically modified (GM) crops, organic farming, free-range chickens or pigs, problems faced by families on very low incomes and any issues relevant to your local area.

Do a web search for 'sustainable farming' — the www.sustainweb.org website might be useful.

Test yourself

20 What are the advantages of:
a grow-your-own vegetables?
b keeping chickens?

Homework questions

1 Which definition best describes sustainable development?

- **a** Development that provides for present needs but may change the environment
- **b** Development that provides for future needs by changing the environment
- **c** Development that has low environmental impact and provides for present and future needs
- **d** Development that has high environmental impact and provides for present and future needs

2 As part of the policy for sustainable development, the government proposes 'a reduction in development on "green-field sites" and increased use of "brown-field sites" '.

- **a** What is meant by the term 'sustainable development'?
- **b** How will this proposal benefit the environment?

3
- **a** Is biohol the answer to solving fossil fuel problems?
- **b** What social problems would biohol production cause in a developing country?
- **c** What environmental problems might biohol production cause in the UK?

4 Incinerators have a bad press and many of the problems associated with them are now controlled. List problems that are managed and are no longer a concern.

5 What problems remain with the use of incinerators to dispose of household waste?

6 In the past, sportsmen fishing for bass often landed fish of over 4 kg and these carnivores gave the anglers a good fight. Nowadays, bass of this size are far less commonly caught and many are only around 2 kg. Commercially, in the UK bass are caught on baited lines, but continental fishing vessels use paired trawlers. In the UK there is a minimum landing size of 36 cm (this minimum is soon to be increased).

Bass spawn in warmer waters and then migrate north to UK estuaries for a few years to grow; they are protected from fishing in these estuaries. Dolphins often pursue shoals of bass as they migrate north.

- **a** How does having a minimum size for landing fish benefit the fish stock?
- **b** What does a reduction in size of bass caught by the sports fisherman suggest?
- **c** Suggest three environmental impacts of the use of paired trawlers.
- **d** Much of the bass sold in restaurants comes from farmed sources in the Mediterranean. What conditions here result in rapid growth?
- **e** Suggest three reasons why the public might be concerned about the environmental impact of fish farming.

7 Go to the following website:
www.guardian.co.uk/environment/2008/jun/11/greenbuilding.food

Read the article 'Welcome to Thanet Earth: is this a taste of future for UK agriculture?'

- **a** **(i)** Record all the ways in which this enterprise could be considered 'green'.
  (ii) Record all the possible objections.
- **b** Explain the possible roles of (i) bees and (ii) parasitic wasps.

Exam corner

Deforestation and setting up a cattle ranch both increase greenhouse gases in the atmosphere.

a Explain three ways that this occurs. *(3 marks)*

b Why do small village clearings cause less damage to the ecosystem than deforestation? *(3 marks)*

Student A

a Carbon dioxide is increased by burning the spare branches of the trees cut down. ✓

The cattle are ruminants and so produce methane as they break down the grass anaerobically. ✓

There are no trees to photosynthesise. ✗

b Small village clearings are abandoned after a few years and so seeds of plants can blow onto the earth and grow. ✓ As they are small, animals can cross the village to move to new areas if they need to find more food or mates. ✓ Small clearings do not change the climate or rainfall pattern. ✓

Examiner comment (a) Student A has given two good points with an excellent explanation of methane production but the third point does not relate to 'increase'. No plants means no photosynthesis — correct, but this does not increase the amount of carbon dioxide.

(b) Three sound points each addressing different aspects and emphasising the importance of the word 'small'.

Student B

a The bacteria in the soil break down the tree bits and produce carbon dioxide quickly. ✓ Cows produce methane and carbon dioxide.

b Deforestation uses big machinery that compacts the soil, so no new plants can grow. ✓ The large patches of bare soil heat up in the day and cool down quickly at night — this causes clouds that might give heavy rain and cause erosion. ✓

Examiner comment (a) The first point about bacteria does state that carbon dioxide is the product of the breakdown of the tree remains. The point about the cows does not mention *how* they produce methane and carbon dioxide so does not get a mark because there is no explanation.

(b) There are obvious signs of some recall here. The first point is worth a mark because a reason why plants can't grow is given. The second point explains the climate change so 1 mark is given. There is not enough for the third mark.

Note the standard expected in an explanation and the use of technical terms such as anaerobic, soil compaction etc. In part (b) a comparison is required and neither candidate has really addressed this.

Exam questions

Chapter 9

1 A chip shop owner claimed that if he soaked his potato chips in water before he fried them, they were firmer than if they were cut up and fried without soaking them.

a Explain how scientific ideas can be used to back up his claim. Use correct scientific words in your explanation. *(4 marks)*

When some uncooked chips were soaking, a worker in the shop poured some salt into the water and left the chips soaking in salty water.

b Suggest how this may have affected the potato chips. *(2 marks)*

A student carried out an experiment to see if adding salt to the water that potato chips were soaking in affected how much they changed in mass while soaking. She soaked one potato chip in pure water and one in a concentrated salt solution. She measured their masses before and after they had been soaked.

c (i) State the dependent variable in this investigation. *(1 mark)*

(ii) State the independent variable in this investigation. *(1 mark)*

d Do you think that the results from this investigation would provide enough information to draw a firm conclusion about the effect of salt concentration on the change in mass of potato chips? Explain the reasons for your answer. *(3 marks)*

2 Oxygen is absorbed from the air into the blood in the lungs.

a What is the name of the process by which oxygen is absorbed into the blood? *(1 mark)*

b Explain how the contraction of the rib muscles and diaphragm aid inhalation. *(3 marks)*

c Describe two features of the lungs that make them very effective at absorbing oxygen. *(2 marks)*

d Explain how the arrangement of blood vessels in the lungs ensures that oxygen is constantly absorbed into the blood. *(2 marks)*

3 a What is the function of villi in the small intestine? *(1 mark)*

b The concentration of glucose in the blood surrounding the small intestine is often much higher than the concentration of glucose in the small intestine itself. However, glucose is still absorbed into the blood under these conditions. Explain how this takes place. *(2 marks)*

c People who suffer from coeliac disease have villi that are much shorter than normal. They often suffer from weight loss and a lack of energy. Explain why coeliac sufferers have these symptoms. *(3 marks)*

Chapter 10

1 a Explain how the blood passes from the heart to the lungs and back. Name each blood vessel and chamber of the heart that it passes through. *(2 marks)*

b What two changes occur to the blood in the lungs? *(2 marks)*

A patient gets breathless walking and is told that she needs a new heart valve.

c What is the function of a valve in the heart? *(1 mark)*

Another patient was rushed into hospital after suffering severe chest pains.

d What causes the pains? *(1 mark)*

e How could this be treated in hospital without a major operation? *(1 mark)*

2 A plant is put in a pot with a radioactive isotope in the water. The route taken by water in the plant is followed.

a Explain in detail how the isotope reaches the leaves from the soil in the pot. *(4 marks)*

A few days later the isotope is traced to the developing plum fruit.

b Explain the form in which the isotope reached the fruit and through what vessels it passed. *(3 marks)*

3 A sprinter is about to run a 200 m race. His starting pulse rate is 60 beats per minute.

a At the end of the race, the sprinter's heart rate has increased to 130 beats per minute. Give three reasons why his heart rate increased. *(3 marks)*

b The volume of blood pumped per beat increased from 90 cm^3 at the start to 150 cm^3 at the end of the race. Calculate the increase in cardiac output per minute. Use this formula:

$$\text{cardiac output} = \frac{\text{volume}}{\text{per beat}} \times \frac{\text{number of beats}}{\text{per minute}}$$

Show all your working. *(3 marks)*

c Draw a sketch diagram of a vein in the leg and label it clearly to explain why blood does not flow backwards to the feet when a person stands up. *(2 marks)*

Chapter 11

1 A person who suffers from kidney failure may be treated either with a dialysis machine or with a kidney transplant.

a Explain how a kidney dialysis machine works. *(3 marks)*

b Give one advantage and one disadvantage of dialysis treatment. *(2 marks)*

c Give two advantages of a transplant as the treatment for kidney failure. *(2 marks)*

d A person receiving a kidney transplant has to take immuno-suppressant drugs for the rest of their life. Explain why this is necessary. *(3 marks)*

2 Body temperature is monitored in two ways.

a Describe these. *(2 marks)*

b Describe how heat is lost through the skin as a response to an increase in body temperature as a result of exercise. *(3 marks)*

c On cold days, people tend to produce more urine, even though they drink a normal volume of fluid. How is the volume of urine that is produced controlled on a cold day? *(3 marks)*

3 A type 2 diabetic teenager is told that he must learn to regulate his blood sugar level by diet and exercise. He is advised to eat protein with wholemeal bread at breakfast because this will be broken down slowly.

a What effect will this have on release of insulin? *(1 mark)*

b He is advised to avoid high-sugar drinks. Explain why. *(2 marks)*

c He is advised to walk increasing distances each day until his weight begins to reduce. How will walking have an effect on his blood glucose level? *(2 marks)*

d Medication in the form of Metformin was used. Suggest two results of this treatment on his blood. *(2 marks)*

Chapter 12

1 Biogas production takes place in the UK on farms and in municipal waste composting.

a What is the typical composition of biogas? *(1 mark)*

b Why is the heat produced from the combustion of biogas less than that from the combustion of methane? *(1 mark)*

c State two advantages of the use of biogas generators on farms. *(2 marks)*

d State two advantages of biowaste digesters in urban situations. *(2 marks)*

e Explain the effect on gas production when the surrounding temperature falls? *(2 marks)*

2 Over-fishing has taken place in the North Sea and there are few full-sized fish caught.

a What impact will this have on future fish stocks? *(1 mark)*

The aim of imposing fishing quotas was to try to develop sustainable fishing.

b What is meant by 'sustainable fishing'? *(2 mark)*

c What problems are associated with 'quotas'? *(1 mark)*

d Explain the problems associated with intensive fish-farming. *(3 marks)*

e What are the advantages of fish-farming? *(2 marks)*

3 a Garden centres are springing up all over the UK and horticulture is a big industry. The composted waste from municipal composting schemes can be used in growbags to replace peat. Why must the material be pasteurised first? *(1 mark)*

b Carbon dioxide is said to be 'locked-up' for many years in trees and peat. Explain what this statement means. *(2 marks)*

c Biogas and biohol are described as carbon-neutral. Explain the term 'carbon-neutral'. *(2 marks)*

d You fill your car with one of the new fuels that is 85% biohol. Apart from the 15% petrol, suggest reasons why the biohol component is not totally carbon-neutral. *(2 marks)*

Index

Note: Page numbers in **bold** indicate where definitions of key terms will be found